Woods and Wells

DESIGNING YOUR NATURAL HOUSE

Preface by architect David Wright

JOHN WILEY & SONS, INC.

New York Chichester Weinheim Brisbane Singapore Toronto

This book is printed on acid-free paper. ♾

Published by John Wiley & Sons, Inc.
Published simultaneously in Canada

Library of Congress Cataloging-in-Publication Data:

Woods, Charles G.
Designing your natural house: over 200 rules of good architecture you can apply to new or remodeled work / Charles G. Woods and Malcolm Wells : preface by David Wright.
p. cm.
Includes bibliographical references and index.
ISBN 0-471-28528-5
1. Usonian houses. 2. Architecture—Environmental aspects.
I. Wells, Malcolm. II. Title.
NA 7117.5.W66 1992
728.37—dc20 92-14954

Printed in the United States of America

10 9 8 7 6

Other books by Charles and Mac:

None.

Other books by Charles G. Woods:

Natural Architecture
The Complete Earth Sheltered House
(Illustrated by Mac)

Other books by Malcolm Wells:

Underground Designs
Energy Essays
Gentle Architecture
Notes from the Energy Underground
Classic Architectural Birdhouses and Feeders
Underground Buildings
The Successful Contractor
How to Build An Underground House
Sandcastles

Contents

About the author, Charles G. Woods (1953 -)

An award-winning designer, registered AIA architect, and author of Natural Architecture and The Complete Earth-Sheltered House, Woods's work has appeared in 50 magazines and newspapers. He has lectured widely and is the junior partner of Natural Architecture, which is headed by registered architect John J. Martin.

Woods studied under a direct student of Frank Lloyd Wright, and Wright is the main influence on his work. Woods also studies philosophy and lives in an earth-sheltered house in the Pocono Mountains of Pennsylvania with his wife, Julie, who is also an author.

Charles is happy to work on natural architecture with interested clients.

Charles G. Woods dedicates this book to his teacher, master architect Dennis Blair.

I'm happy to say that in the eight years I've known Charles I've seen him become more and more committed to environmental issues. —Mac.

Malcolm Wells (1926-)

I'm the person who penned all the lines and illustrations in this book but most of the words and ideas, and, I suppose, most of the designs, came from Charles. Surprisingly, even though I have known him by phone and letter for about 8 years, and have both admired his books and worked a bit on them, I have never met Charles Woods. We plan, when our first royalty check arrives, to meet somewhere halfway between his eastern Pennsylvania and my Cape Cod, to see if our friendship can stand the strain of a face-to-face encounter. New Haven looks like a reasonable spot.

I wandered aimlessly during the first 40 years of my life, blaming everyone else for the land destruction and the ugliness that were even then changing this beautiful continent. When I finally saw what I'd been doing as an architect I decided to specialize in underground architecture.

Now, some 25 years later, I do consulting design work in, and write books about, that slender branch of architectria. My dream is of a continent made green again, put back as close as possible to the way it was, with all of our manworks underground, looking out upon it all from the sunny rooms.

I guess I should explain, while I have your attention, that I'm the "Mac" mentioned throughout these pages, and it is I to whom Charles . . .

Acknowledgments

Charles again:

I would first like to thank editor Larry Erickson at Better Homes & Gardens for suggesting an article on this subject. Also Malcolm Wells for suggesting that my article be expanded to book length, and for adding all the wonderful drawings — and his long experience — to this book. In my opinion, he is one of the greatest... (Hey, Charles, I can't write all this garbage about myself. What? OK, OK.) ... architect/environmentalist/illustrators of all time!

I would also like to thank Editor Wendy Lochner of Van Nostrand Reinhold, and her assistant, Kelly Francis, for having faith in us and our project.

I would also like to thank especially my wife, Julie K. Gundlach, for proofreading, and my whole family, especially my aunts and my father-in-law, Herman Gundlach, for their continued support.

I would also like to thank the folks at Mail Market, the Printery and Duff Maps for their help.

Finally, I would like to thank my architecture teachers chronologically: Mr. Eperson and Mr. Phil Knudsen in high school, then architect Dennis Blair (a student of Frank Lloyd Wright), my teacher for over ten years. Then architect Albert Sincavage, another

...has graciously given his permission to say anything at all in these bottom strips.

Wright student, then "unofficially" Malcolm... (Charles! No! No more!) Wells and my senior partner John Martin. Also master builder Larry Wilson, and of course through his work, Frank Lloyd Wright, my main influence.

Lastly, friends and architects I have learned from in various ways: architect David Wright, my Elder Brother (thanks for all the time on the phone – and the preface). Jay and Tracy Boyle for a lot of support and a review of this book. And also architects Don Watson, Dan Lawrence, Lester Wertheimer, my brother Michael, and other influences too numerous to mention.

I would lastly like to thank my clients, especially the Stultzes, Frattinis, Wollenburgs, and Angellos.

Oh, almost forgot! Karen North Wells for fielding <u>a lot</u> of phone calls to Mac!

<u>And now Malcolm:</u>

There is a number of people I'd like to thank for all the help and encouragement they've given me. I'm thinking particularly of 3, 7, 10, 23, and of course, 9. Then 6, 14, and 5. And a special thank-you to number 18 for that glorious night in Buffalo.

PREFACE *By David Wright, Environmental Architect*

Mac and Charles have collaborated to create a primer of good and simple architecture — a book like this. It has the technical clarity and hand-done quality of Frank Ching's books (V.N.R.); it has the friendliness and humor of Forrest Wilson's books (V.N.R.), yet it has a broader, more general, public appeal. This book should reach out to people who are designing or having a home designed for the first time. It has many technical and aesthetic points poignantly portrayed as "rules". Most of these rules are based on a combination of practicality, efficiency, concern for beauty, and an over-riding love for well thought-out and aesthetically-pleasing architecture.

Malcolm (Mac) has a unique, classic eye for architecture and ability to sketch. His genius is topped off by a relentless sense of humor — a most barbed wit that always strikes a vein.

Charles' text creates a pattern of ideas, opinions, and descriptions that express technical concerns and design techniques as a logical part of an integrated holistic architecture. The co-author doesn't always agree and that helps draw the reader into the dialog.

These guys are architects' architects (of the organic Mr. Natural School), putting their combined experience, common sense, pet peeves, and tips into a highly readable, graphically delightful (sometimes contradictory) compendium of architectural rules.

I believe this book of wonderful sketches, useful design ideas, pervasive humor and philosophy of architecture will reach out to a broad range of readers. I would not be surprised to see:

- **Architects** buy this book for potential clients (and as gifts for other architects!)

- **Teachers** recommend it for students of drawing, design and architecture.

- **Couples** purchase it along with "plan" and "idea" books when they start thinking about designing a home.

- **Professional custom builders** use the ideas for spec houses and to stimulate their potential customers.

In short, I think this is a unique, original, and positive addition to the library of books about architecture.

It might remain timely as long as people build houses the "old fashioned way." Because it is general enough, it addresses aesthetics, energy conservation, practical concerns, order, economics and most other aspects of the architecture design process.

I believe this book is worth the price for the humor and drawings alone!

Sincerely,

David Wright, AIA
Environmental Architect

Introduction

We cannot teach you, in 200-odd pages, how to design houses like those done by the great architects. (It would take at least 350 for that!) But almost all architects agree as to what is wrong with the typical "builder's-type house". In this book we have tried to give you some of the do's and don'ts of basic good house design.

Malcolm and I are within the organic/modern tradition and have been very much influenced by the work of Frank Lloyd Wright... We do believe, however, that most, not all, of our "basics" hold true as well for modern, post-modern, and even traditional/Colonial design.

...Although you'd never know it to see my work
MW →

There are of course exceptions to almost all these do's and don'ts. Many have been disregarded by geniuses. (Most have been disregarded by everyone.) Even we have varied from them at times. But it's usually best not to disregard most of them on the average house. In any case, we believe you will be benefited by your reading of this book.

Malcolm and I have added several complete designs of

ours (as well as some by others) at the back of the book (Don't peek!), giving the reader examples that illustrate many of our <u>do's</u> and <u>don'ts</u>.

Please remember to build modestly; we're destroying the earth's resources. Use wood sparingly, use no tropical hardwoods, redwoods (or Charles Woods!).

We hope you'll enjoy our ideas but don't take them too seriously. It is easier to say what is bad than what is great in architecture — as in life.

Good reading!

—Charles (and Mac).

P.S. I want to apologize for some of Mac's humor, which isn't funny. I have tried to change him but have failed miserably. At least I bear the brunt of most of his dagger thrusts.

Don't forget: Earth second, too. Man third. And: "Less is more". —Mies Van Der Rohe

Priorities

This book is mostly about the way buildings look. It's the kind of thing you'd expect from two building designers. But the appearance of a house is almost trivial compared to the physical impact it has upon this battered land of ours.

A building is a very destructive thing, not only during its construction but during its entire existence. That's why we've tried to remind you – <u>and</u> ourselves – on each page that the real issues go far beyond those of proportion, taste, color, texture, and style. A desperate worldwide environmental crisis is at hand, and every decision we make has some sort of impact upon it.

Here's hoping this book will help us all begin to change that impact from negative to positive.

Charles G. Woods, "Planet Earth, 1·1·92"
and
Malcolm Wells

SAVE THE WHALES ...and everything else.

A Philosophy of An Organic Architecture by Charles G. Woods.

Do we need a philosophy of architecture? That question is not a prelude to an intellectual game; it is the very foundation upon which the work of an architect or designer rests. Are we lacking a coherent philosophy of architecture? My answer: Just look around!

Look at our cities: 100-story towers of steel and glass, dehumanizing in scale, loom as clones of one another. Hubris characterizes their demeanor. The glass on all four sides mockingly attests to the lack of concern for energy conservation. There is defiance, not deference, toward nature.

Witness the chaos represented by our gas stations, our fast-food restaurants, our shopping malls, and even our houses. We see a mix of Greek, Roman, Mediterranean, and contemporary design elements, sometimes in the same structure. There is no proportion, no sense of wholeness, no sense of relationship to nature, to neighbors, to history. We are drowning in an architectural sea of forms and details. We need a raft – a philosophy of architecture based on consistent principles.

I do not presume my own work to represent the ideal. But in this book I have tried to design buildings based on a set of principles. Using a term applied to the works

Go get 'em, Charles!

Philosophy (cont.)

of Wright and Sullivan (those great and lonely geniuses who almost single-handedly established a natural and rational American architecture), I characterize my designs as "organic" in the sense of 'living' and 'whole' with interdependent parts; for that is how I visualize a building. Architecture creates for humankind a second skin. It is revealing that the word "house" often symbolizes "body" in the Bible and other spiritual and religious writings.

Organic architecture is mystical as well as rational, humble as well as inspired. Its principles are wholeness, simplicity, and honesty. Organic architecture must take into an account an array of very human considerations. Besides cost and practicability, the key to living, holistic design is the environment – its geography, topology, history, climate, and color and materials.

Let us begin our look at organic architecture with the concept of wholeness, for as the foundation supports the structure of the building, so the concept of wholeness supports the philosophy of organic architecture.

The idea of nature as an organism, in a sense of a living body (panpsychism), has a long history. It is evident in the writings of the pre-Socratic philosophers, Heraclitus

Whew! No wonder I was a failure as a teacher. I just think of designs and then draw them. —M.W.

Philosophy (cont.)

and Parmenides. In the Renaissance its champion was Spinoza, later, Hegel and Darwin. Today Nature as a living organism is seen in the developing philosophies of Samuel Alexander, A.N. Whitehead, Sri Aurobindo, and Teilhard de Chardin.

All of nature is one organic whole. Everything is interrelated in a dynamic way. Humanity (at least the physical body, for those who believe in a separate and immortal soul) is a part of nature. Although many have tried, humanity cannot be cleaved from and set in opposition to Nature.

We pollute Nature both physically and visually as if it is not in any way related to ourselves; whereas, in fact, it _is_ ourselves. This havoc we wreak upon Nature has its origins in religious and metaphysical error. Unconsciously we ascribe to the dualism of Rene Descartes, whereby humanity is separated from Nature. This concept, with its corresponding hubris and ignorance of what humanity truly is or can become, manifests itself in ugliness. We pollute Nature first with the mind – with the concepts and ideas – and only later with toxic wastes and unsightly and dysfunctional buildings.

Architecture is great, just as art is great, when it proceeds from an intuition of the ordered relationships and the wholeness that exists in

Philosophy (cont.)

Nature. Architecture can both reveal the essence of humanity and critique it. It matters not whether one subscribes to the idea of Spinoza and the oriental philosophers that humanity has an essential nature or whether one sides with the Existentialists to declare that humanity has no essential nature other than what it wills or creates itself. Architecture holds a mirror to both these positions.

Some may feel that to impose a philosophy on architecture is to wander far from the path an architect must tread. I think not. The concept of Nature as a living and vital organic whole may be argued, but it cannot be denied. Wholism is revolutionizing medicine, psychiatry, ecology, and more. It will revolutionize architecture. We may view the works of Antonio Gaudi, Rudolf Steiner, Eric Mendelsohn, Le Corbusier, Mies Van Der Rohe, Herman Finsterlin, and, of course Frank Lloyd Wright and Louis Sullivan to see the evolutionary and revolutionary impetus of organic architecture. On the contemporary scene, we may glimpse the future via the works of Malcolm Wells,* Paolo Soleri, Roland Coates, David Wright, etc.

These architects and designers have accepted the obligation and duty of their professions – to

* How'd I get in there?

Philosophy (cont.)

design for the essence of what we are and can become. Architecture, as our reflection, should show us at our best.

Architecture tells us much about ourselves and about our past. What does the architecture of societies in the past represent? Egypt: the striving for eternity. Greece: the pursuit of beauty and grace. Rome: the drive for power. The Gothic cathedrals of Europe: the yearning of finite beings for a transcendent God. It is not happenstance that Hitler and his architect looked to Rome as the model of Third Reich architecture. Gigantic in proportions, scale models of Speer's buildings were sometimes one hundred feet long. These gigantic designs dwarfed human proportions as surely as the government they symbolized effaced human individuality.

We have had great architecture in America. Early colonials developed a beautifully simple architecture in response to their environment, but it was soon bogged down in duplication of ornamentation from the past. Early colonial architecture is superb compared to later gingerbread designs.

(It is interesting to note that in historical societies with a linear view of time, architecture tends to retain historical details from other periods. However, in non-

Philosophy (cont.)

historical societies, the so-called 'primitive' cultures — those with a cyclical view of time — one sees a more perennial architecture. For example, I submit that Roman architecture is not as great as Greek architecture, for the former superimposed the latter over its own innate Roman identity).

And what of America's architecture today? If we were to judge by influence alone, there are four or five unknown designers who would be called great. Representing over 90% of all housing built in the United States, their designs are familiar to us all. But if we were to judge those designs by the use of energy, cost, or aesthetic standards, they would surely be seen as disasters.

We see diamond shapes in entry doors next to French provincial lanterns, next to colonial windows. I do not intend to criticize individual homeowners. Perhaps Americans expect too little of their architects. Nonetheless, their architects have failed them.

The post-modern movement continues this tradition of disaster, creating, for instance, buildings with Ionic columns and fluorescent lights painted pink and blue. These are the most famous architects around! One cannot deny they are brilliant and creative, but their misplaced

Philosophy (cont.)

creative energies serve only to degrade architecture to an indulgent, even silly, individuality.

Happily, there are beautiful exceptions. For instance the architecture of Frank Lloyd Wright was an attempt to create a truly American architecture, a departure from the hodgepodge of Greek and Roman stone details imposed on wood structures. Although Wright had, and continues to have, great influence on American architecture, his philosophy has really never taken root. His Usonia houses are incredibly beautiful. Wright, and Louis Sullivan before him, have much to offer. Why not design according to the principles they represent? That is what I have tried to do in this book. Building on the foundation of the wholeness that exists in Nature, I have used simplicity and honesty as my guiding principles.

Simplicity need not be dull, as the Japanese have shown. In all its periods, Japanese architecture is easily recognizable. Yet it has many variegated forms, and in addition to its basic simplicity it is honest.

What do we mean by honesty in architecture? Let us begin with what honesty is not. It is not asphalt shingles made to look like wood. It is not plastic made to look like stone, or rubber made to look like

Philosophy (cont.)

brick. Wood, brick, and stone are beautiful, but if they are too expensive to use, seek honest alternatives. A concrete slab floor may be scored with module lines or stained a natural color. Remember: Be honest. Do not even try to make it look like brick; it is concrete!

An organic house should virtually design itself. All one must do is ask the right questions and follow the logical conclusions. What is the watertable and how does it affect our design? Can we cover the home with earth, or should we only partially berm? What local building materials are available — stone and hemlock? Conclusion, we build a bermed structure of stone and hemlock.

Where does the sun shine? Where does the wind blow? We are confronted with a rugged site strewn with boulders and dotted with clusters of old oaks. We terrace downhill, placing our home among and around mighty trees and giant rocks.

The organic home is simple. It uses only a few materials, and it uses them honestly. We design the house on a module for structural and design uniformity. Elevations reflect the module in its glass and post placements. The smallest detail reflects the whole, and the whole is related to each part. The house is easy on the eye.

The colors are natural and healing. The spaces are proportioned to the human body; cabinets are designed so that short as well as tall persons can conveniently use them. The organic house is naturally warm in the winter and cool in the summer. It relates compatibly with its setting, both wild and man-made.

All this and more is what I have tried to accomplish in the design and points in this book. I welcome dialogue and discussion on these ideas, and I sincerely hope I have not offended architects and homeowners with these frank remarks. However, I hope I have stirred their thought processes, for what I have written I firmly believe to be the truth.

—Charles G. Woods

This essay was reprinted with permission from Element Books and is a revised version of the original.

Amiens, ca. 1270

Charles asked me to include a drawing of a rose window. He said it was an excellent example of "organic architecture." Having never really looked closely at one, I resisted, so he sent me a whole book* on the subject. I was bowled over, completely unprepared as I was for the beauty and skill exhibited in the work.

When I think that these huge windows were made of <u>stone</u>, over 700 years ago, by people using simple hand tools, and assembled high above the ground, then intricately glazed with richly colored glass, I am amazed.

* <u>Rose Windows</u>, by Painton Cowen; 1979/Thames & Hudson, NY.

De Veritate*

Charles G. Woods again at the podium:

A note about what I don't like about 'post-modernism', 'deconstructionism', and 'traditionalism':

The post-modernists' criticisms of some aspects of the 'International School' of Modern Architecture (not the organic school) was somewhat justified but they threw the baby out with the bathwater!

Mixing historical styles is silly, not Beautiful! (Although I do like some of Philip Johnson's work!)

Deconstructionism reflects to me the chaos of our times. It is eccentric and brilliant but again I do not believe it is beautiful. Possibly it is not meant to be.

Colonial-traditional.

If you want to look as if you live in an old house go and live in one. But to repeatedly copy such design seems to belittle what was great in it. And there was much!

Architecture has seemingly lost all hope of finding the Beautiful, just as our society has also lost almost of hope of finding the True and the Good.

Much of modern architecture, I believe, rests*(philosophically) on nihilism (unbeknownst to many of the architects themselves).

* On Truth.

(* It is influenced by writers like Derrida in philosophy.)

Being underground, I've missed the whole post-modernism movement. Missed deconstructionism, too, as well as De Veritate. It's well insulated down here. —M.W.

I am also a man of my time. I almost despair of finding beauty in the world, my life, or my architecture, but I have glimpsed it, and I do believe it exists.

The Beauty of God is the cause of the Being of all that is.
-St. Dionysus

But Beauty is not the object of making. Beauty is an accident of <u>right</u> making.
- Eric Gill *

* see Gill, Eric; *A Holy Tradition of Working*. Lindisfarne Press.
Also see Maritain, Jacques; *Art and Scholasticism*.

"My best design probably to date!" — Charles G. Woods

Would you believe my father-in-law would not spend the required $2,000,000 to make me architecturally immortal? — CGW

Gundlach Residence (project, C.G.W.)

Floor plan.

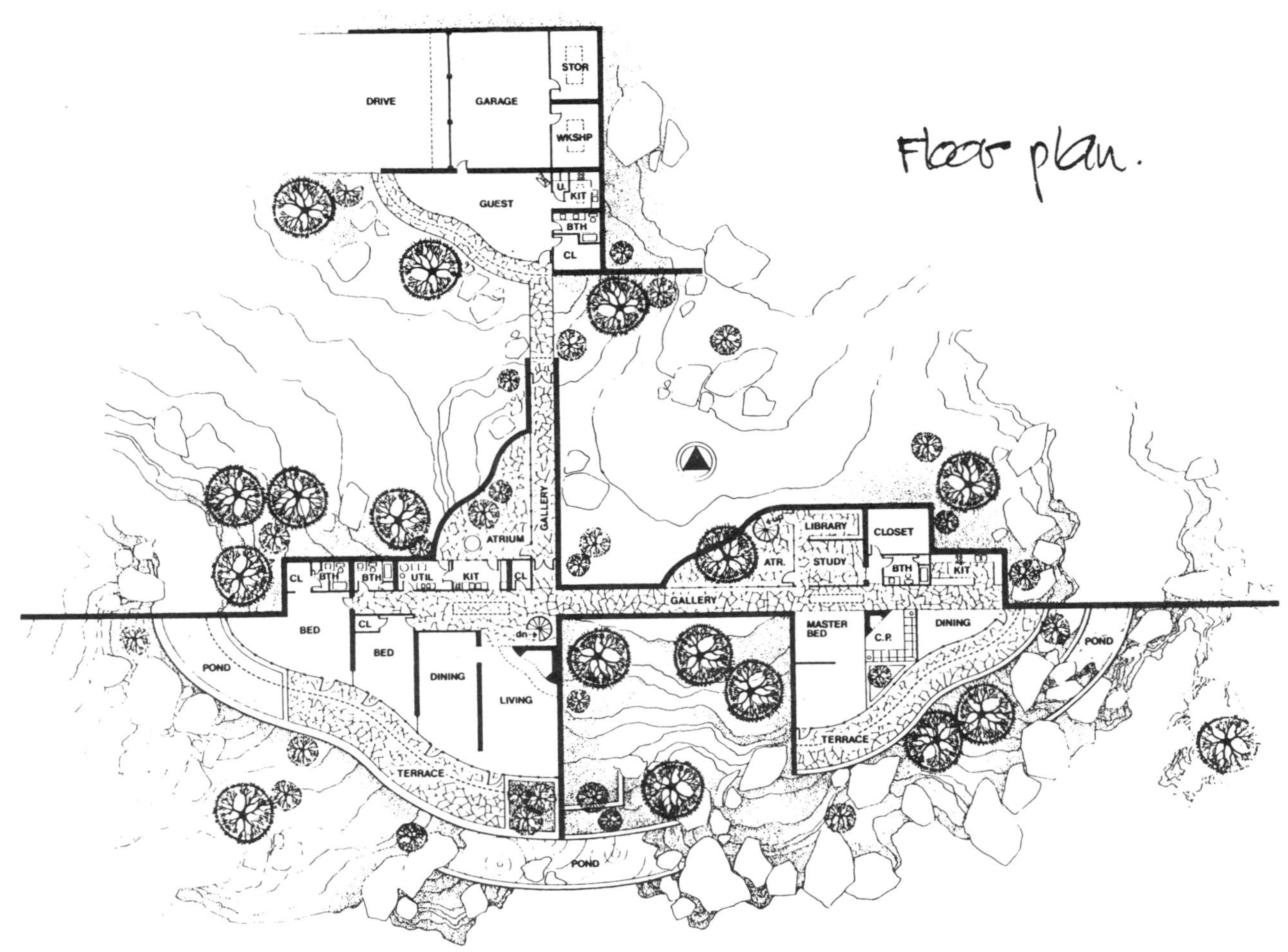

Gundlach Residence

"How much do Mac and I agree/disagree?," asks Charles

And he answers this question, about which practically everyone is wondering, in the following way: 2

Which of us is better? I honestly don't know for sure for Mac is a mysterious man. We have not disagreed often and never harshly. Our two biggest differences, parapets (I: yes, he: no) and boxed-out metal chimney flues (I: yes, he: no), I think are minor in the cosmic scope of things. The spirit of our work is, I believe, similar, and goes back a lot to Wright's work (but Corbu, Mies, Neutra, Aalto, etc. influences no doubt also show).

Charles is far better than I.

Not in mine, I hope. (Mac)

You are arrogant! (CGW)

I think on basics we agree about 90%; on details probably a lot less. I think it's clear, though, what's basic and what's detail in our book.

This book is by both of us but in many ways he's the master. Am I being humble? Let no one misunderstand: I am still a genius.

No, it's the truth: in rendering and detailing and experience (and on larger projects) it's no contest. In design I arrogantly think we are in the same sphere.

Why is he better? Because he's somewhat older and wiser but I do hope to pass him by some decade.

Part of the reason I like Charles is that I see in him a little of myself at that age. Impatient, ambitious, full of ideas. I found old geezers like too slow.

Charles G. Woods on the subject of earth-sheltered housing.

Mac has written more than I, in this book, on underground housing. He is "The Father of Underground Architecture" in my opinion. But I am at least a son, having spent 15 years designing them. Even in high school I did one earth-sheltered house. I have designed over 50 earth-sheltered houses – see my two books.

What is my conclusion? Such buildings are great – but they can be more expensive, and you should not build one without an architect's or an engineer's plans, and never use anything but butyl-type membrane roof – not tar, etc.!

I think that many earth-sheltered houses look like bomb shelters. If you use concrete roofs they must be detailed not only structurally but architecturally. I use mainly wood roofs. I also like a combination of exposed and earth-covered roofs.

If you have the right lot, a good architect, and a good builder, go for it. I live happily in one, and have a wonderful $500,000 house under construction for a client.

I use berms (earth banks) in almost every design.

Malcolm and I disagree on some parts of the earth-shelter

Be sure of your motives when you build. Don't build underground simply to save on heating fuel costs. There are cheaper ways to build that accomplish that.

subject, especially that of parapets. (All right, Mac, I brought it up.) A parapet is a low wall at the edge of a roof, which can be a leak* problem, says Mac, in freezing weather. Soil expands, when frozen, with tremendous force. But I say that a simple camber (slope inside the parapet) solves the problem.

Mac likes ... or even ...

Why do we disagree? Mac's roof looks more simply natural; mine has an architectural line to it. I can compromise, of course; I like both. He forgets that my own earth-sheltered house does not have a parapet.

On and on the playful argument goes, so don't let us stop you from doing either. If you want complete merging into nature, vote Mac; partial merging, vote Woods. Just call 1-800-111-1111.

Mac's heart is in the right place, though! Maybe I'm too arrogant to completely cover up my work.

—C.G.W.

* Leaks of both water and heat often occur at parapets. Even before expansion and contraction weaken them parapets are heat bleeders.

Woodsian interlude. (A good example of curved design. - CGW)

The beautiful thing is that which being seen pleases. - St. Thomas Aquinas.

Solar Arc.

A word or two from the underground man, Malcolm Wells

The more I see of the destruction caused by the paving-over of America with buildings and lawns and asphalt the more certain I am that our architecture is all upside-down. We put the dead surfaces and the machines on top and crush the land to death underneath!

Underground architecture is right-side-up. It puts the living land on top, where it belongs, and puts all the dead things underneath, where they belong. But this book is not about my specialty. It's about all the things we can do to make our buildings look better and perform better until we're ready to go all the way.

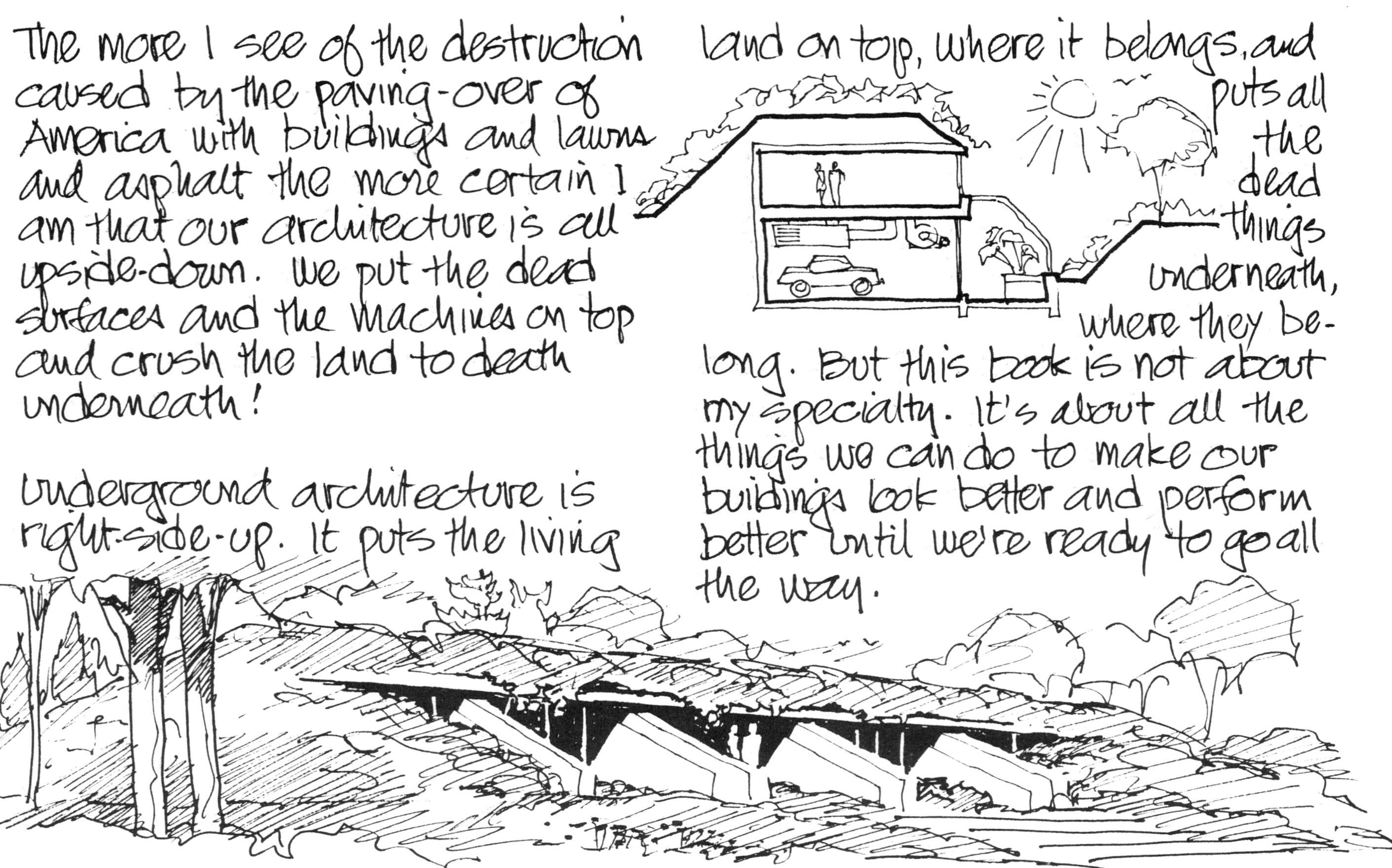

Underground houses are warm and dry, bright and sunny, silent and efficient, cool in summer, and open to garden views. Beautiful.

Be Carful!

Everything about a house, from the beginning of construction straight through its time of occupancy till its eventual demolition, has an element of danger. That's why we have building codes and all the other regulations having to do with safety in buildings. That's why Charles and I have stressed the ideas of planning, of care, and of safety consciousness in various parts of this book.

But we can't cover every eventuality. Neither can all the codes and safety laws. Accidents will happen.

We can only hope that in your case they will be small ones.

So read the labels, follow the directions, get professional help with your designs, plan ahead, build with care, and be on the watch for safety hazards after you move in. We want you to be around for a long, long time.

Whether you build – and live – above or below ground this page is especially for you.

BASIC DESIGN PRINCIPLES

Your house should be in harmony with nature.

e·bon·y and i·vor·y

charles will now tell you a little more precisely what that means:

"This whole book is an attempt to see that you do precisely that."

Aha.

For uncounted eons our ancestors had few or no possessions ...no houses, no clothes, nothing. And then we changed – explosively!

Make isolated features part of the design.

Sometimes a lonely window in a big wall is appealing, but not too often. The usual condition is simply bad design, or thoughtlessness, with unappetizing results.

That's why Charles wants you to add some wood trim and to locate your windows with enough care that the trim lines can make them enrich the design. Not a bad motive. He illustrates the treatment with two explanatory wall elevations; we'll call them one and two.

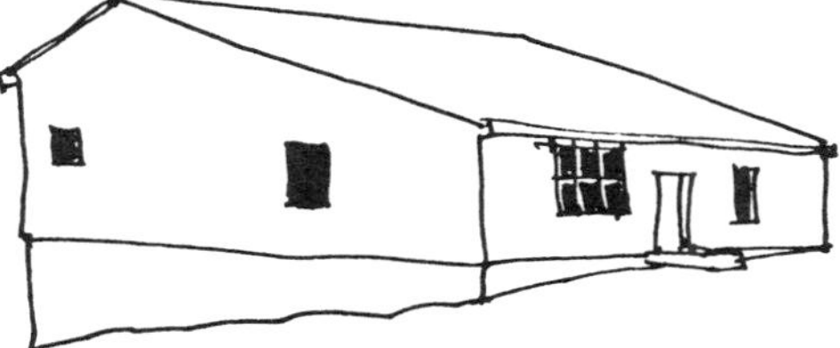

This trim detail emphasizes rectangles, not holes in walls

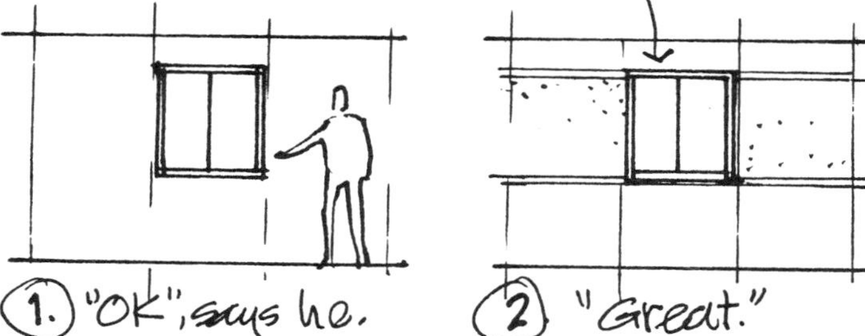

(1.) "OK", says he. (2) "Great."

See that? He's done it again, and with a modular grid to boot.

I'm not as griddy as Charles. I'm too old to learn. I tie things together with a high shelf.

Is not often that a natural scene is dominated by one distracting element. The integration out there is awesome. Will we ever match it?

Sheds, mail boxes, dog houses should exactly match house.

Except in size, he must mean. Otherwise how would you tell which one was the house?

He's a stickler for continuity, all right, but I must admit that everything is related.

Maybe, however, there's a less strict interpretation of his rule possible, one in which some of the spirit of the main design – and materials – is carried through to all the parts.

As you can see, we're trying to nudge American residential architecture an inch or two in the direction of unified organic earth-relatedness

Use the same siding indoors and out, if possible.

"But," adds Charles, "be sure to use interior stain indoors, as exterior stain is toxic; use same color in and out."

And remember two things: 1) be sure all indoor-to-outdoor surfaces and materials have a thermal break (an insulated gap to slow heat loss) and 2) weathering will darken the outdoor wood faster than the indoor, so consider darkening the interior stain to compensate. But then the room may get too dark.

So you'd better

It's not easy, life, is it?

These glassy walls are shown to illustrate the indoor-outdoor look. Remember however that such walls must be carefully designed to minimize heat loss.

Remember the physically handicapped.

Each of us could be in their shoes someday, and if future laws require barrier-free access to all buildings, including houses, before they're sold, you could be ahead of the game by doing it now.

The natural world is under increasing physical handicaps as we become ever more removed from it by our electronic titillations.

Cantilever!

Almost every ounce of wood in a tree is acting as a cantilever. Gravity (G) exerts tons of force on the outstretched arms, wind (W) bends the earth-held trunk. And yet we almost always use wood in simple spans rather than in the cantilevers to which it was accustomed. An overhang, a deck, a projecting beam — each has the spirit of the tree about it. If they made tree limbs the way we make beams each one would have a prop under it.

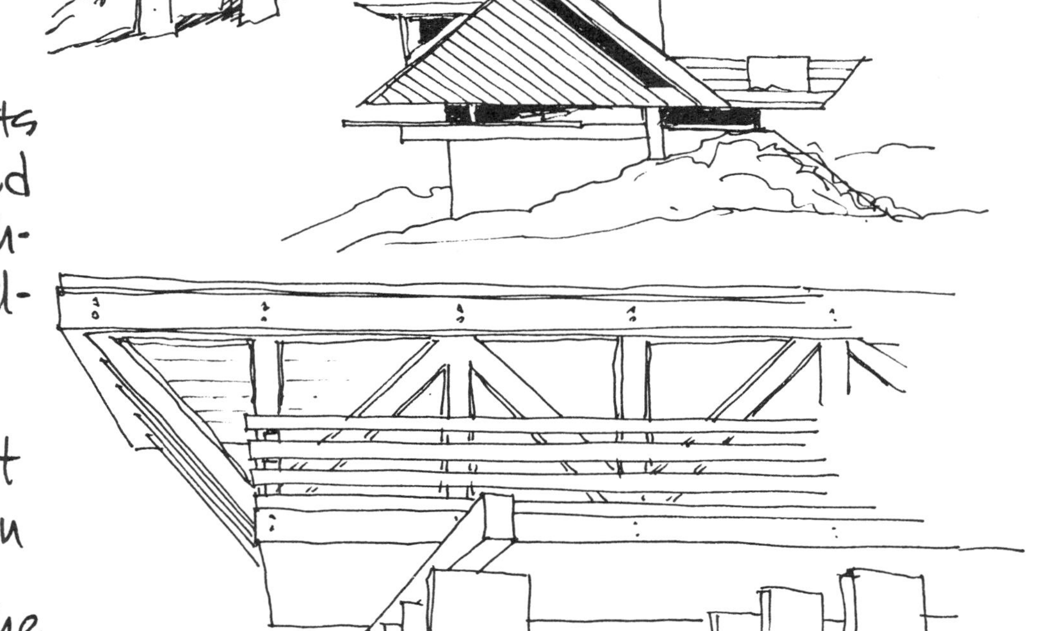

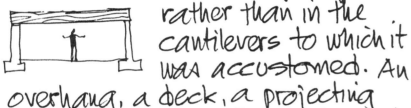

Some cantilevers droop in time so give them some upward camber to start.

Charles says, "cantilever 1/4 distance. One third is possible with a good engineer!"

A good stiff engineer, I'd say, if he's going to cantilever that far.
In any case, design it to last for hundreds of years. True patriotism.

Use fewer columns...

...is the way Charles starts this page, continuing, "and make them larger if necessary to compensate. Remember to use cantilevers, and make decks lower to the ground than the usual inhuman (inhumane) 8 feet. Or make the distance to ground much more than 8 feet.

and line up
those sliders!

If you need headroom under the deck, OK, but put a berm around the lower patio.

Charles says he likes it.

It would take quite a bit of concrete and steel to support a deck cantilevered that far. It's shown here not so much as a suggested design as an illustration of an architectural theme carried out fully, the deck both a continuation of the house and an abstract version of the surrounding trees.

A gentle ruggedness, as opposed to a spindly ricketiness, seems to be more at home in the natural world, which is the only world we have.

The Charles G. Woods Basic House Construction Details sheet.

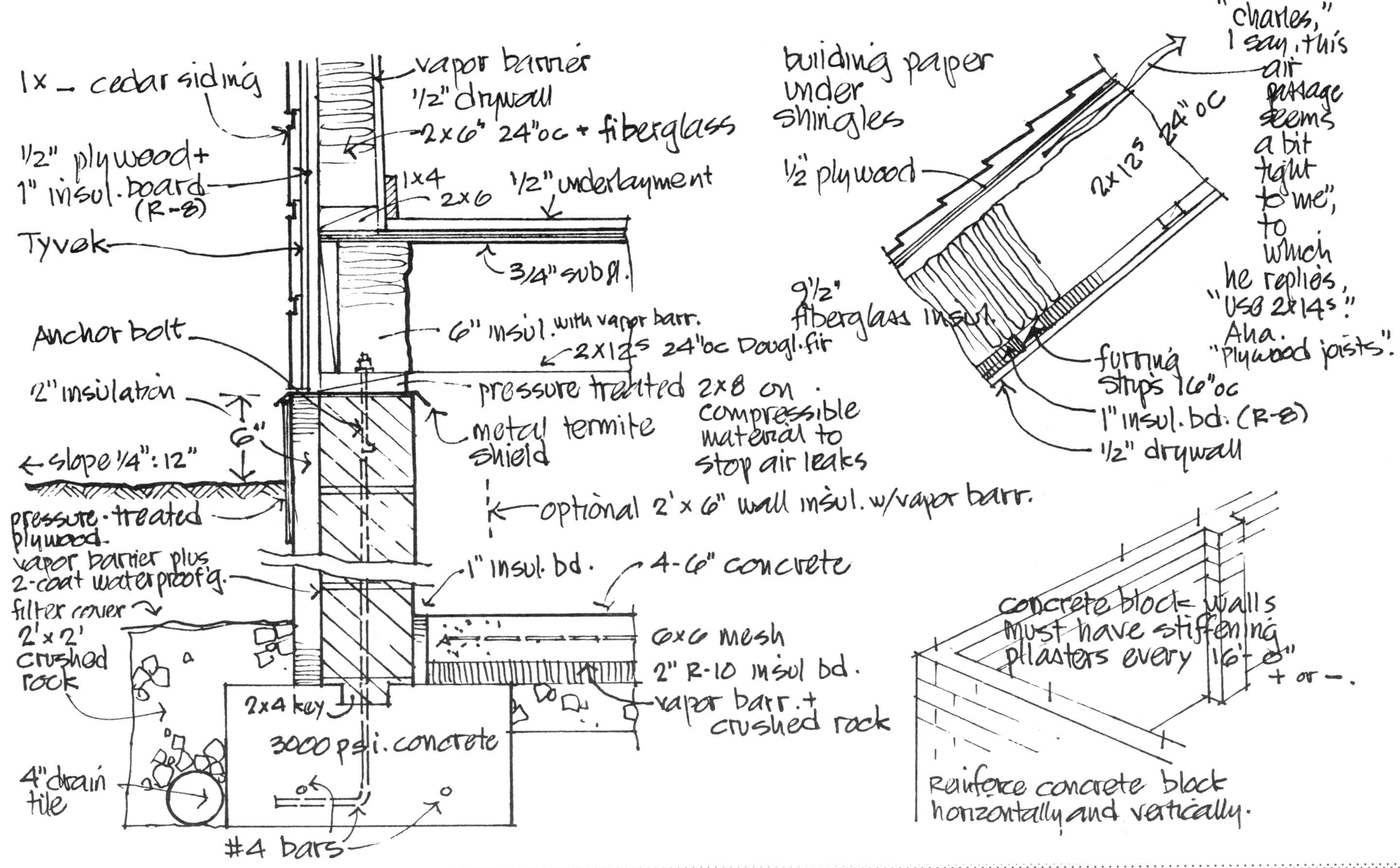

Shredded junk-mail makes poor wall insulation. It's inflammable, too. Maybe our congress will slow the daily tide of waste. Maybe.

Use built-in planters indoors and out.

Things to remember:

1) Match the design spirit of the house.
2) If the planter is indoors, remember to water the plants.
3) If the planter is out of doors, remember to water the plants during dry spells, and to design the planter so that neither ice nor rot will destroy it before its appointed round. Freezing earth expands and will burst almost any container. Wood that is constantly wetted and dried will quickly rot. Don't use treated wood indoors... or for any food-producing plant containers.

plants in pots can be rearranged and rotated for sunlight

pebble bed in metal planter-trough keeps pots from being saturated

pebbles under trough

soil and peatmoss mixture holds water and doesn't freeze as rock-hard as soil alone

stone

sloped sides may encourage earth to rise rather than burst wall when freezing

reinf. bars

If zoos are jails for animals, what are indoor planters for plants? Should they be left free in the tropics? What about food plants?

Consider emphasizing the structure.*

Sometimes, of course, it just isn't practical to leave all the supporting members exposed, so this rule doesn't apply to every house.

But when a repeating structural element is obviously a basic part of the design the structure should be emphasized. It gives the building a backbone and it gives the design meaning. This begins to get more difficult as we wrap our buildings in ever thicker layers of insulation but it's still possible to do it effectively.

*especially on plank and beam construction.

In warm climates, Charles reminds us, you can expose block masonry - and concrete - with nice surface treatments.

Does a clam emphasize the structure of its shell? Does an egg? A leaf? Natural systems are so sophisticated it's hard to say, sometimes.

For a custom house, bring it alive with open plans, fireplaces, skylights, decks, high ceilings, sunken floors, open kitchens, master baths, and spiral stairs. And, if you're competent to handle them, related angles and curves as well.

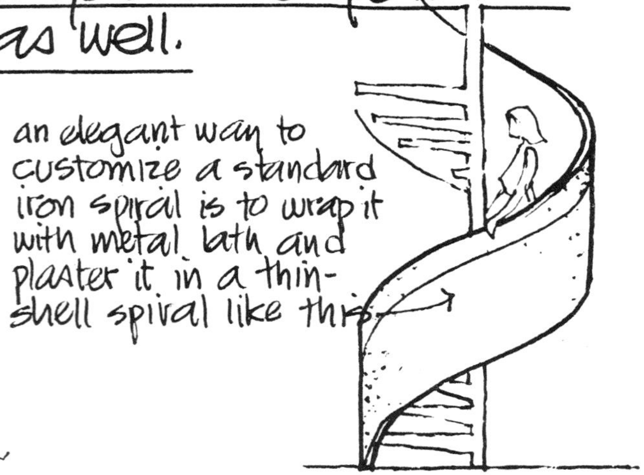

an elegant way to customize a standard iron spiral is to wrap it with metal lath and plaster it in a thin-shell spiral like this

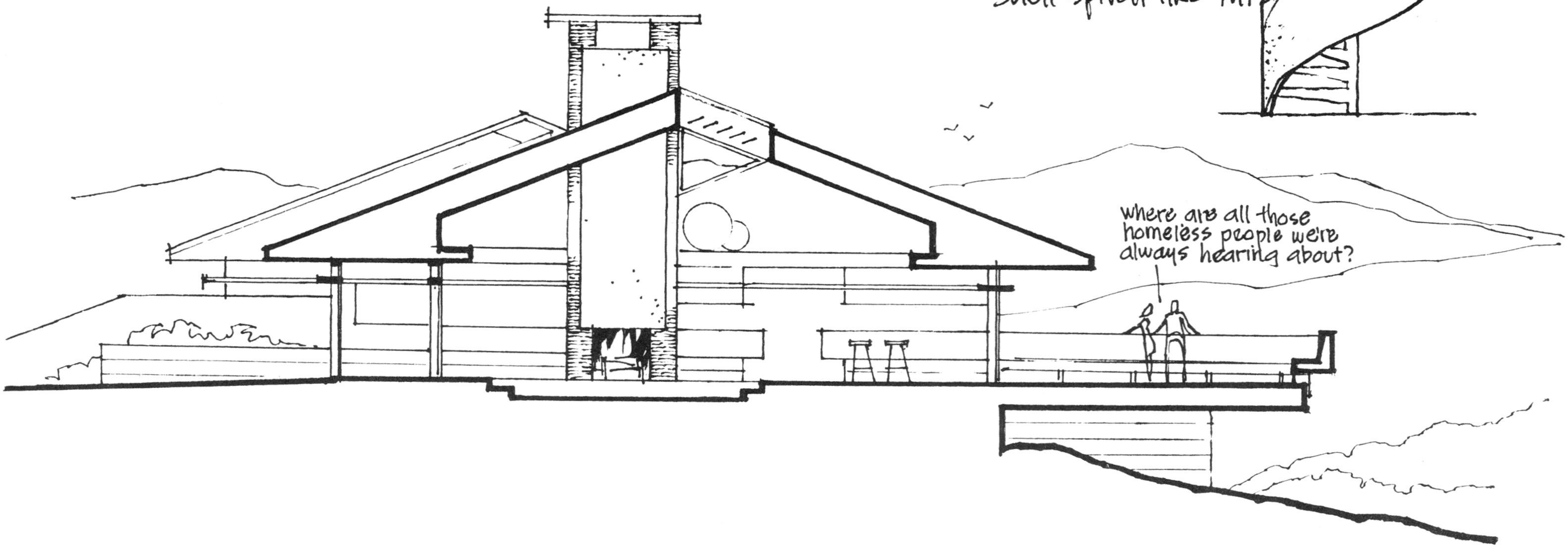

where are all those homeless people we're always hearing about?

It's hard for many of us Americans to realize that most of the earth's human families live in one or two tiny rooms. What will happen when *our* wealth dries up?

"Don't walk through one room to get to another.

Trailers have to, houses don't" –CGW

Use a hall or passageway. It won't be a space-waster if you use it as a storage area too... or an art gallery, a sun space, or a work room.

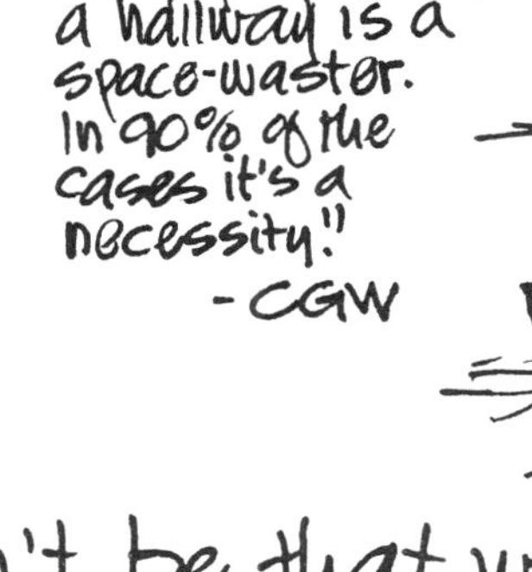
"Many people say a hallway is a space-waster. In 90% of the cases it's a necessity!" –CGW

Too many halls are dark, unimaginative and boring but they needn't be that way...

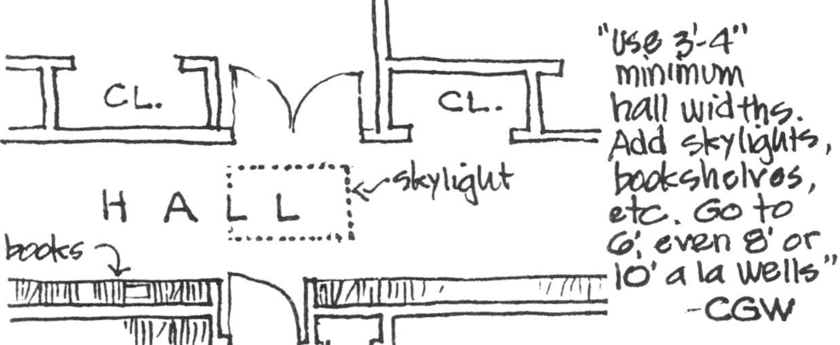

"Use 3'-4" minimum hall widths. Add skylights, bookshelves, etc. Go to 6', even 8' or 10' a la Wells" –CGW

The more uses you can think of for a space like this the more its existence can be justified.

Now that you've got your nice multi-use hall or passageway you can concentrate on making your house ever more efficient and benign.

Play off opposites in your house and landscape design.

High / low.
Inside / outside.
Light / dark.
Simple / complex.
Woods / Wells.
...you get the idea.

Keep 'em guessing what's just around the corner by leaking a hint of it before you get there.

Didn't we already do this one? Oh, well...

"Well," responds Charles, "a little redundancy never hurt anyone."

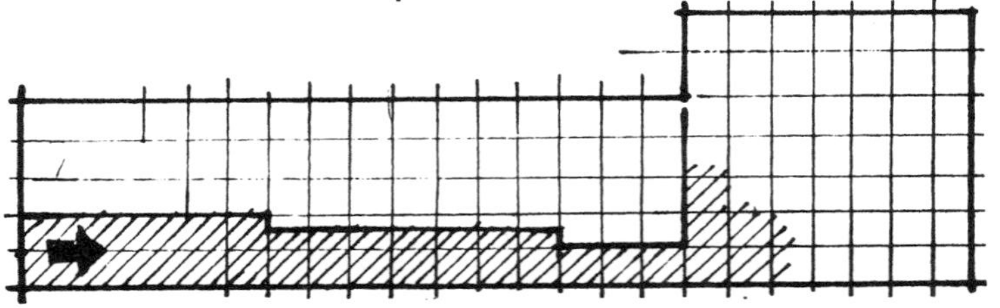

"Take this plan," says he, "with a hallway going from 8' wide to 6' to 4', and then opening into a large room with a high ceiling."

Opposites of type and size can be very effective in a landscape. Just be sure those opposites are appropriate to your region.

Most houses are healthy. It's the occupants who are not.

What you want is a health*ful* house, one that will keep you healthy.

Look around you. Is there a surface in any direction that is not made of chemicals?

Down below you is the vinyl carpet or the asphalt tile. The furniture is full of plastic foam. Its wood surfaces are covered with plastic varnish, your walls with acrylic latex. Ceiling: ditto. Your drapes: polyester. Blinds: plastic. The windows, at least, are made of good, old-fashioned glass. Down in the basement, radioactive gas called radon may be escaping from the earth into your house. The flames of your gas range are polluting the air, and electromagnetic radiation is zapping you from your t.v. set, your computer, your electric blanket, your clock, and the nearby power lines outside. If someone in the house is a smoker the chances... but you know all that. The question is how do you feel? Have your kids been well? What about lead paint residues? Asbestos? Is your food free of pesticides? Do you spray poisons throughout the house to control fleas? Moths? Ants? Are you OK?

This is a good place to mention The Natural House Book. For information on it, see page 88.

Use restraint with your ornamentation.

Sometimes, a restrained line of dentils will add a nice touch of movement to a frieze board but it doesn't mean

you should hang those tulip shutters just because Dad got a jigsaw for Christmas. By the time he's done all 16 windows and the mailbox it could start to look a bit much.

Many materials - wood, stone, brick, polished brass, weathered copper, even rusty iron - are ornament in themselves when used simply and naturally. It's not necessary to improve on Mother Nature. House designs can be bold if the boldness follows the general theme of the building but intentional distractions can only weaken the design and make Charles and me feel ill.

"The house itself is the ornament."
—Charles

When you cut back on your mowing you'll have more natural ornamentation than ever; wildflowers will thrive, and - cut - they look great on the table.

Stick to one design theme inside the house, too.

Bath and kitchen cabinets, for instance, should seem to belong to the same house. These two don't:

Does it matter?
After all, they both work perfectly well. The answer is no, it doesn't matter as long as you're willing to wear a sock of a different color on each foot. It's that kind of a situation; not earth-shaking but embarrassing when you're able to see how funny it looks. And it's far less satisfying than doing it right.
Now, how about these? Don't they seem more architectural, less like big boxes someone lugged into the rooms?

Floor tile, hardware, ...everything matches.

It seems to follow that if you are sensitive to these design issues you'll be sensitive to all design issues, and the environment will benefit from it.

Forgetting complexity and contradiction—and genius—for a moment, a house should at least reflect wholeness, simplicity and honesty.

(Only senators and congressmen may build dishonest houses.)

simple and honest but lacking something.
You're tough sometimes, Mac.
CGW

The beginnings of wholeness, simplicity, and honesty. You can see, almost at a glance, how all the parts relate.

Simplicity not only relates to design. It relates to saving money, and it makes landscaping easier, more appealing, and less expensive.

Unless your house is done in authentic traditional style,

Don't use old-fashioned lamps and lanterns.

Even if your house is almost traditional get rid of those fakey eagle-topped aluminum electric versions of the old oil and gas lanterns.

As we've explained so beautifully on the previous page, light sources should never be seen. They blind the viewer to the things he should see in the night. (Worst of all, of course, are the outdoor flood-lights that shine directly outward from the house into the eyes of the visitor.)

Ideally, we should light objects, not display lighting fixtures, and there are many ways to accomplish this.

And always use the new energy-saver light bulbs. Fluorescent, they last up to 10x longer.

Darkness – the quality of true nighttime – is now almost unknown except in the most remote areas. It's a shame our kids must miss it.

All foursides of the house count.

Charles says this is "real" important.

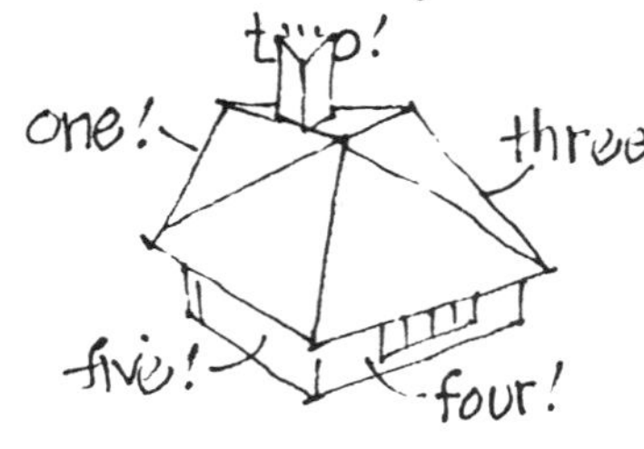

... at least, that is, if the house happens to have four sides. In any case, the point is that there shouldn't be one "feature" side, all dressed up to show itself off. And yet we see owners, builders, even highly-respected architects, producing false-front monstrosities that will embarrass us as long as they stand.

Too many of us say "what's the big deal? It's only the side (or the rear) of the house."

Next time you're invited to Buckingham Palace would you be willing to dress this way? If your answer is yes go back to page one and start over. If your answer is no then don't send a poor innocent house out onto the street with a false front, either. This isn't kindergarten any more. No more make-believe! Our buildings as well as our lives must be honest. Honest.

False fronts, just like all the "green" advertising now so popular among the big corporations, must finally give way to telling the truth.

Make the house speak clearly

Houses can't talk.

Oh, yes they can. And most of them send garbled messages. How many times have you walked up to a house and found yourself unable to decide which door to enter?

Now *that's* ambiguity.

What is this guy trying to say? "I'm a wooden house," or "I'm a stone house"? "Brick"?

And this one: is it a western ranch or an Olde English tavern?

Does it matter if we produce architectural garbage? Of course it does. It means that *we're* ambiguous as well; uncommitted, valueless.

Let your creations speak with honesty and simplicity, and then let's move on; we have a lot of important work to do.

Do not mix historical styles.

(At least not unless you get special permission from the authors. Send $100 in unmarked bills for license.)

Parthenwater

Hey! Why not?

Houses should be organic,

not so you can eat them without having to worry about pesticides but so that all their parts are comfortabely related to each other and to their sites. Flat country calls for horizontal emphasis, cliff faces for vertical accents, and urban sites for good neighbors.

Don't fight the site.

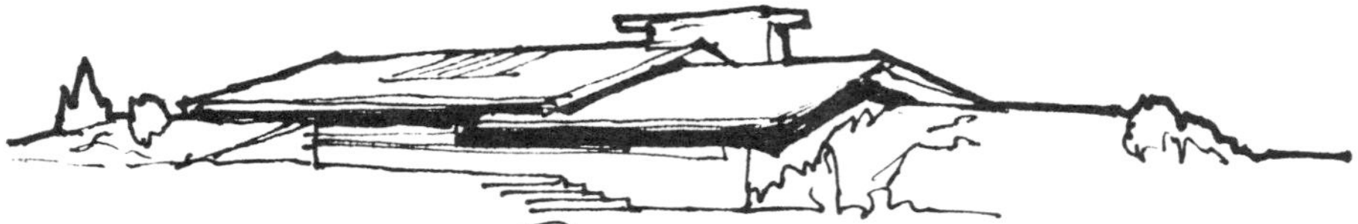

Wanted: good neighbors in my human family.

— Mother Nature

Massing and scale, while hard to define, must be considered.

Sometimes it's easy to tell when a house is ill proportioned...

...and sometimes it's not. You just get a feeling that something is wrong here.

So: how can you be sure that <u>your</u> house passes the test? Easy: just send me a certified check in the amount of one thous...

<u>Malcolm</u>! Stop it! No!

I was just kidding, Charles. Look, to prove it I'll draw some good proportions anyone can use. OK?

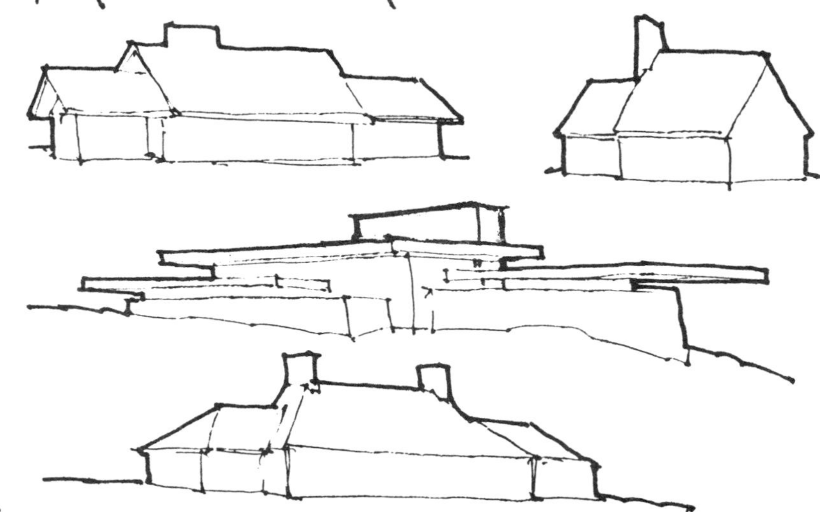

If you want to see perfect proportions, massing, and scale, all you need do is step out of doors and look at something green.

Do not use circles, curves or angles in your floor plan unless you really know what you are doing.

And if you do use them be sure they're reflected in the exterior design.

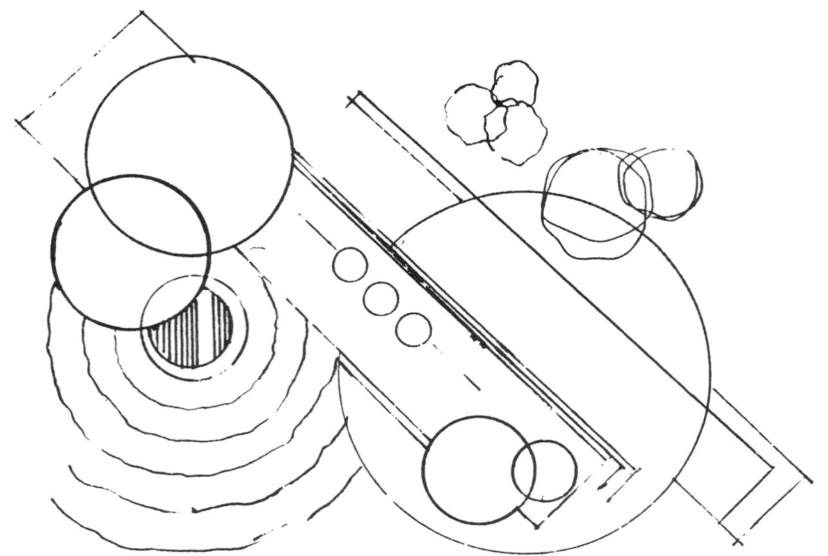

Pick your theme and then stay with it.

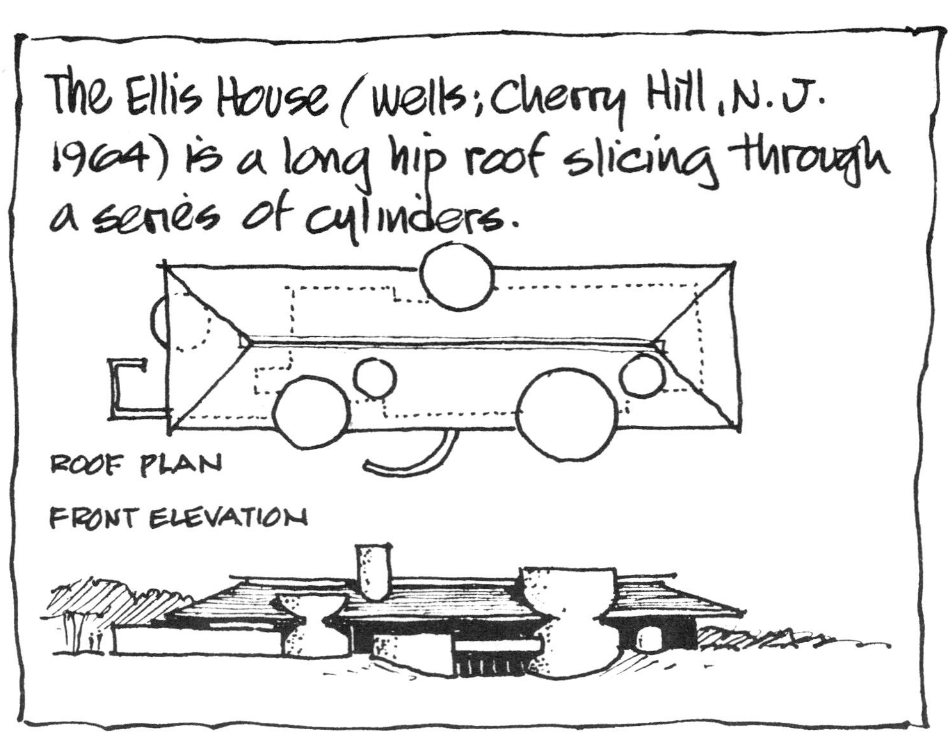

The Ellis House (Wells; Cherry Hill, N.J. 1964) is a long hip roof slicing through a series of cylinders.
ROOF PLAN
FRONT ELEVATION

For further information on this subject go to a natural area near you and think of the ways circles were used in that masterpiece.

Watch out for unintegrated ells on plans and on roofs.

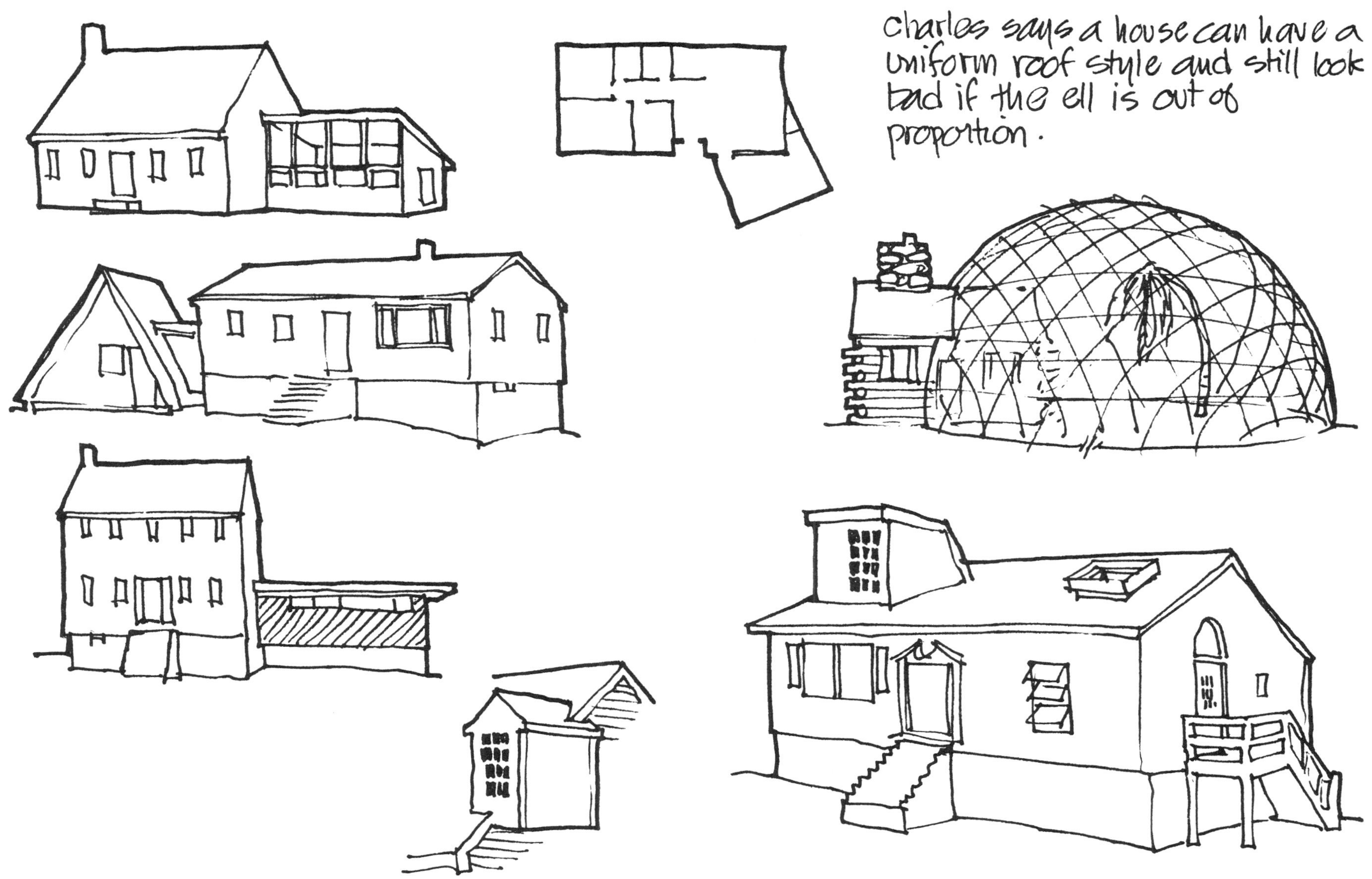

Charles says a house can have a uniform roof style and still look bad if the ell is out of proportion.

After all that we humans have done to poor old Mother Nature, this is the last straw: making her sick looking at our monstrosities on her land.

Don't overbuild*. Save natural resources and money. Spend some of the money saved on good design, better quality work.

We invite you to fill in the environmental moral of this page: ____________

Consistency and Expression: don't forget them!

"For instance," Charles says, "I saw a nice modern colonial house but a couple of skylights wrecked it. You might get away with a small one but if you put in a large modern roof window it clashes."

This reads "contemporary"
This reads "traditional"

I want to be high

I want to be low to the ground.

Here Charles illustrates the idea of expression: letting each element be fully what it is. Not compromising.

"Actually," adds he, "an even sharper contrast would be even better."

American housing has for decades been consistent in its wastefulness and ugly in its expression.

A compendium of actual houses we've seen. across America

Is that the new freeway running under the basketball court? This must be what they call eleganto.

Yes, it's true: too many cooks do spoil the broth. The poor thing is too embarrassed for words.

People call it the hangover till we point out all its features.

Four generations of us Mozebys has lived here and we're proud to call it home.

A veritable fortress of stone (veneer), and a solid feeling of safety ... until that walk gets wet! Nice landscaping, folks.

These 11 guys combine to waste more energy than perhaps 50 well-designed

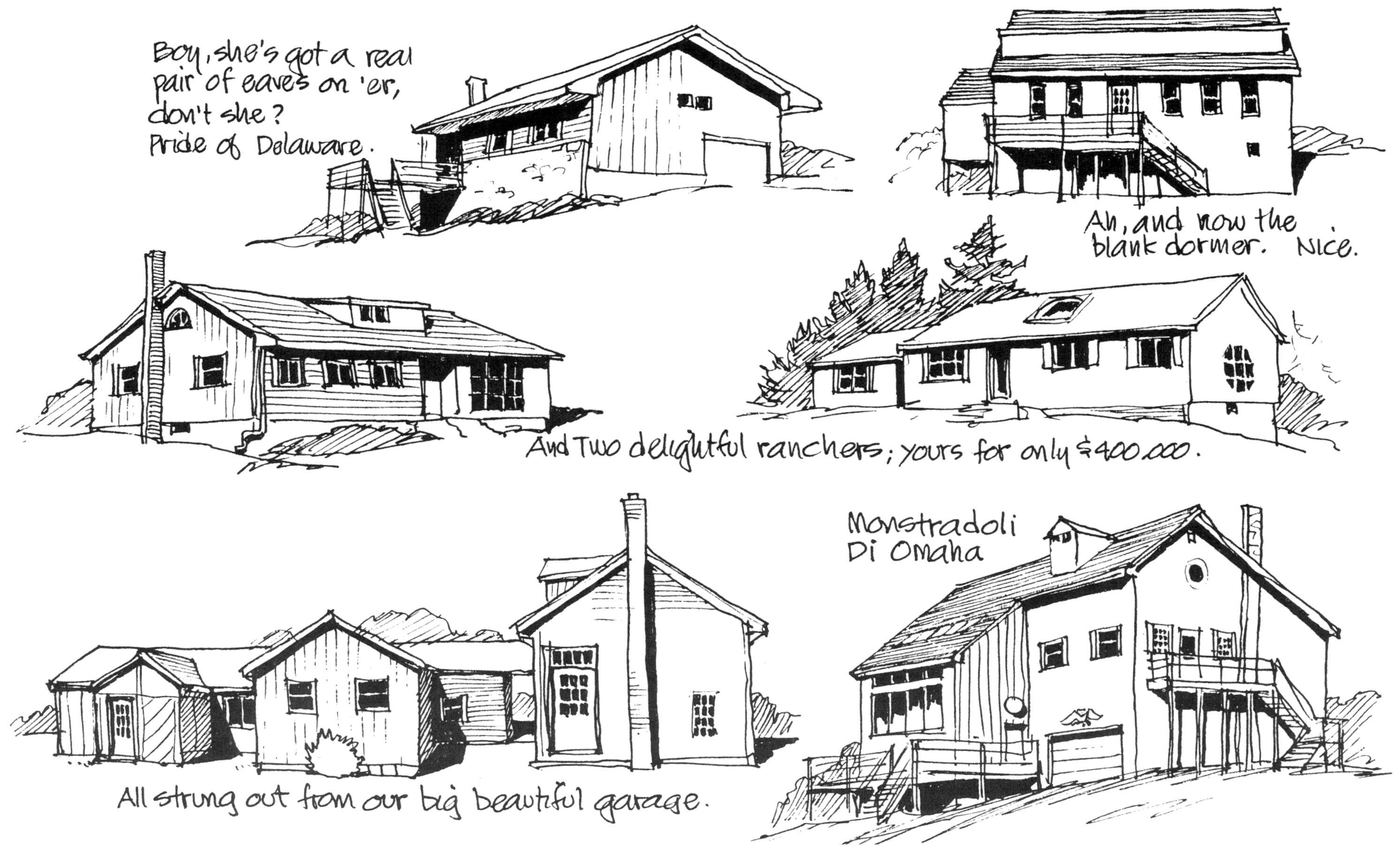

houses would. Retrofitting them will be expensive.

Boy, she's got a real
pair of eaves on 'er,
don't she?
Pride of Delaware.
Ah, and now the
blank dormer. Nice.
And Two delightful ranchers; yours for only $400,000.
Monstradoli
Di Omaha
All strung out from our big beautiful garage.

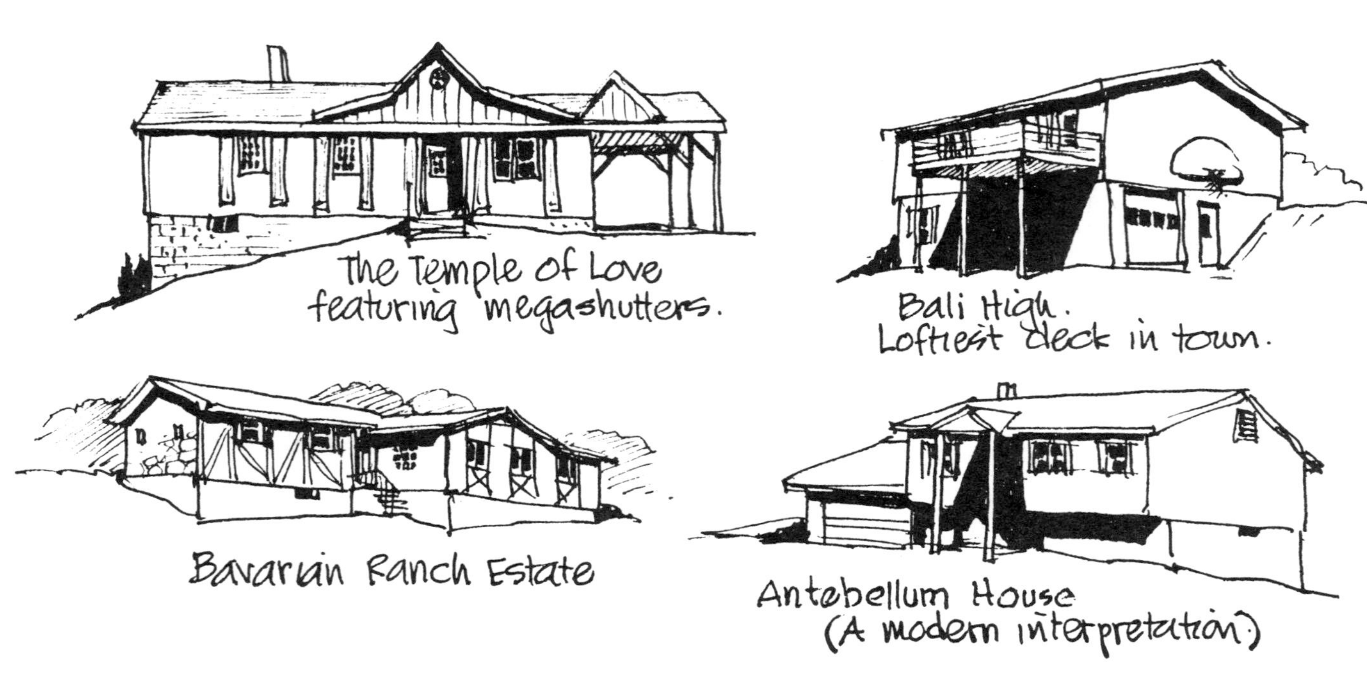
The Temple of Love
featuring megashutters.
Bali High.
Loftiest deck in town.
Bavarian Ranch Estate
Antebellum House
(A modern interpretation)

Olde
Maine Street
"We couldn't
find replacement windows
tall enough so we filled
in the openings with
vinyl siding."

Merge the inside of the house with the outside.

In the bad old days of modern architecture we were so good at getting the "indoor-outdoor effect" people were always smashing into the glass, unaware of it, and the heat from the rooms poured away through those continuous planes of floor, walls, and ceiling like an energy nose-bleed. Awful. It was architecture without thought for the consequences.

Nowadays we can have the same effect without the damage, first by preventing accidental smash-through, and second by thermally separating the indoor and outdoor surfaces. Here's that same room done today:

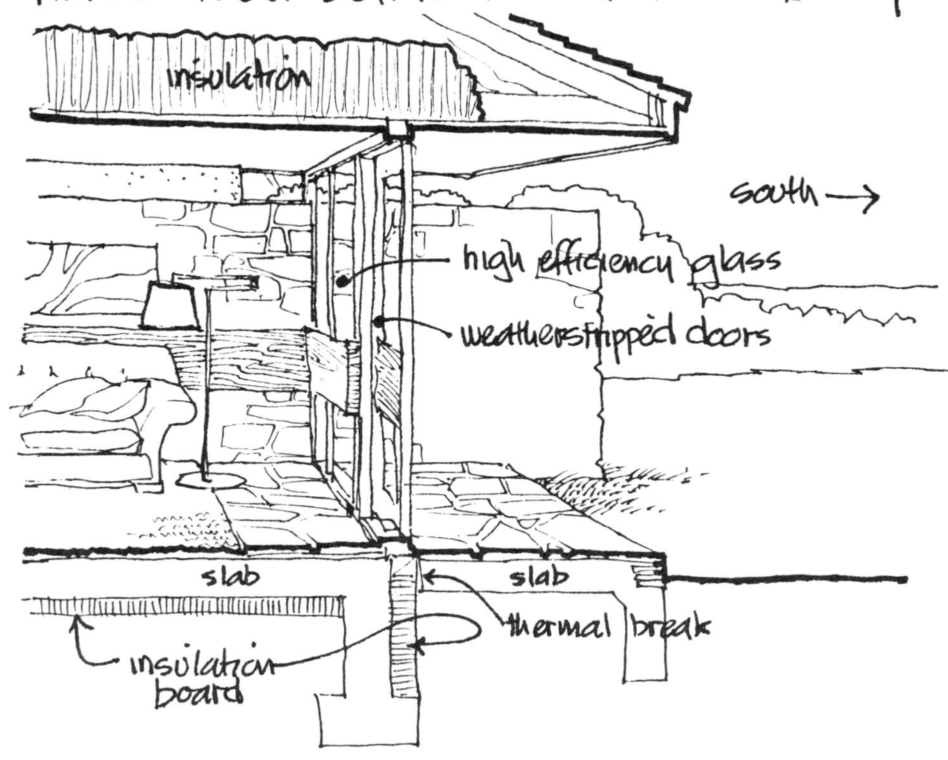

"Task lighting" – putting the light exactly where it's needed – has come a long way, too. (50 watts 2 feet away = 200 watts 4 feet away.)

Vary ceiling heights

The only thing worse than one of those ceiling lights smack in the middle of the room is to see that the dreary ceiling (at its 7'-6" or 8'-0" inhuman height) is likely to extend at that same height throughout the entire house.

Houses shouldn't be cluttered with too many planes and surfaces but they are supposed to be interesting. If the ceilings first compress and then expand the space you will almost feel, as well as see, the architecture as you move through it.

The drama would be heightened if the floor went down as the ceiling went up but that would show an insensitivity to the handicapped*.

*Charles has asked me to show how a 12-foot ramp could be added over a 12" height change. He adds that small spaces can be boring.

Lower ceilings help to enhance small spaces. Small spaces conserve energy, materials, and natural resources, and you save $ on fuel.

Consider using atria (open or covered), courtyards, and greenhouses.

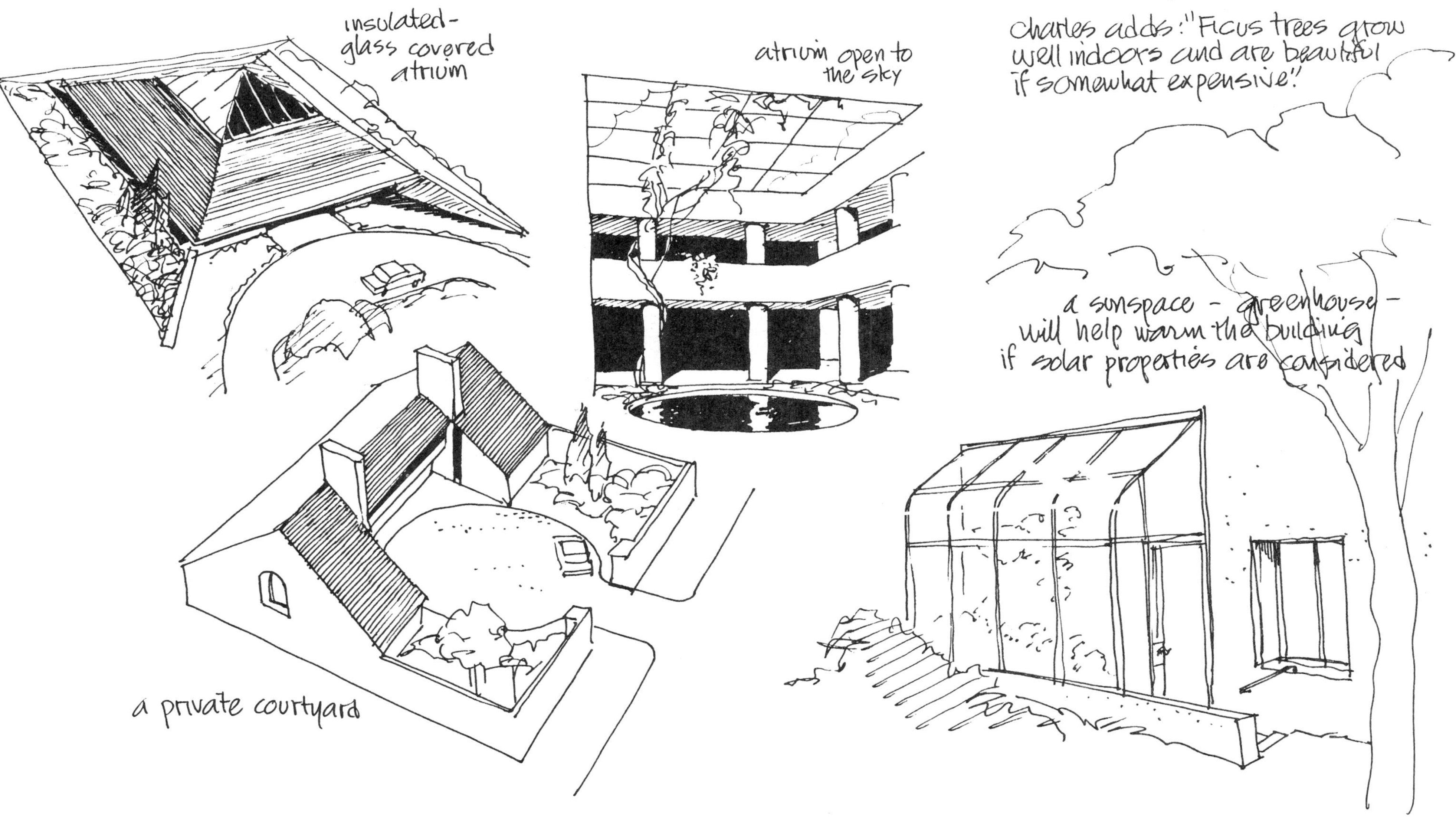
insulated-glass covered atrium
atrium open to the sky
Charles adds: "Ficus trees grow well indoors and are beautiful if somewhat expensive."
a sunspace - greenhouse - will help warm the building if solar properties are considered
a private courtyard

Don't let an atrium or sunspace create air-conditioning problems. Use shade or natural ventilation as needed to prevent heat build-up. see Stultz residence in the plan section.

Stick to simple, consistent shapes in preparing your floor plans.

I thought this said ILUOTAC when Charles first drew it for me but now that I understand it I heartily agree with him. Too many good designs are hampered, even ruined, by letting the floor plan run away with the designer.

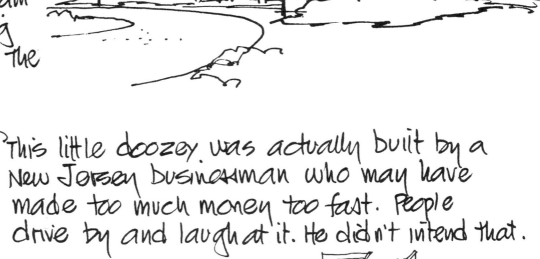

This little doozey was actually built by a New Jersey businessman who may have made too much money too fast. People drive by and laugh at it. He didn't intend that.

Rooftop view.

An integrated form would have been much better.

It's not a crime to engage in gross material-waste but there is a penalty. It's in the air we breathe, in the water, on the beach.

simple, consistent shapes (con't)

"If you want to impress people impress them with good design and your taste in whom you choose as your architect. Houses like that N.J. clunker usually have anonymous architects. Build a little smaller and hire a top-notch designer. The fee? Perhaps 10 to 12%. Then you will impress people and have a good house." Charles continues: "Here are some more basic plan shapes."

If it's close to square make it square, not 28' x 30'

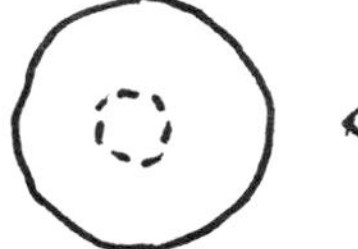

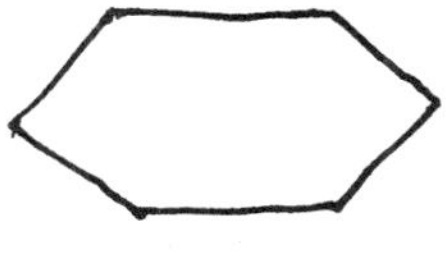

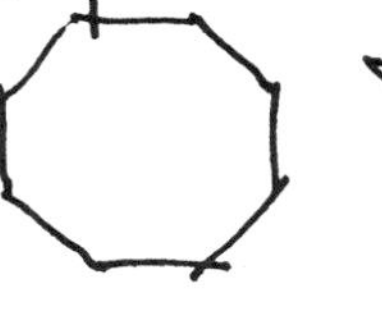

"I have suggested using 4' square modules but you can also use rectangles or angled shapes as the modules. Even degrees of arc can be used as did my guru, Frank Lloyd Wright. You can also use large modules in conjunction with the basic grid."

or, more expensive (a la Aalto):

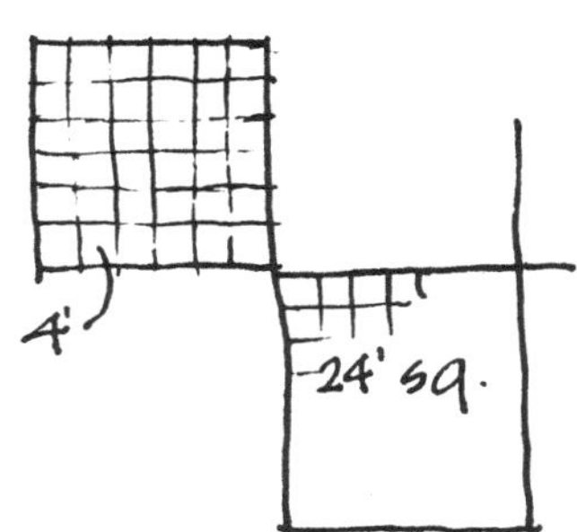

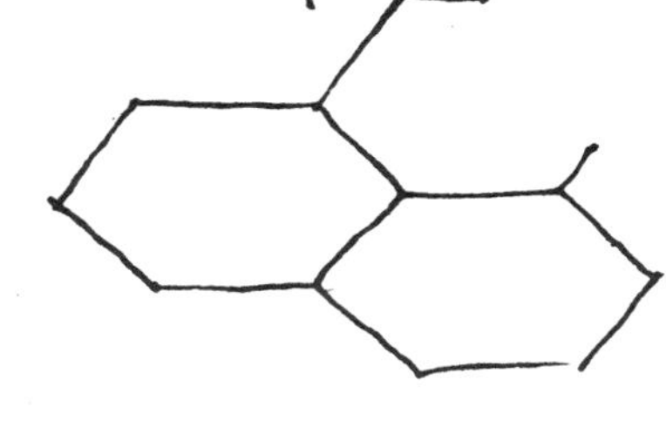

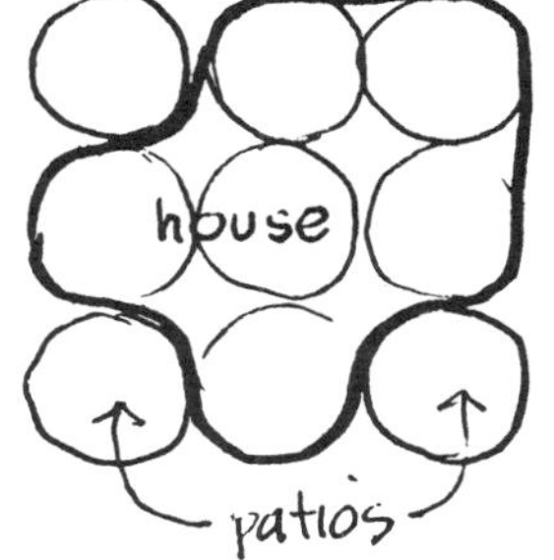

That Charles! What'll he think of next?

Speaking of incredibly beautiful geometric constructions, flower gardens are astounding in that regard. Consistent, efficient, colorful, miraculous.

Conceal satellite dishes

Better?

Even out here
Charles is still
trying to phone me..

On the other hand,
if you happen to <u>live</u> in a
dish-shaped structure...

...but he finds
the line is busy.

"bzz.. bzz... bzz.
"Ha ha ha!"

Have you read Jerry Mander's book, <u>Four Arguments for the Abolition of Television</u>? It makes you think twice about our being fed by tube.

No visible T.V. antenna ... or even utility poles!

That Charles is merciless, isn't he?

What are we supposed to do, go without power, telephone, T.V., fax, and modem? (Hmm...doesn't sound like a bad idea, now that I think of it, but that's not what he means.) Hide the antenna; put it in a tree. Put it in your attic, but hide it.

OK, Charles, but you can't get the power company to move a pole that happens to stand in front of your masterpiece.

"At times you can!", says he.

What spirit that man has! Next thing you know he'll be against letting sewage flow in the gutters.

There's only one thing wrong with hiding poles, wires, and antennas: their quiet takeover of our lives has no reminder. Out of sight, out of mind.

From Charles the wing wall man:

Wing walls extend the lines of a house, and help the indoor-outdoor effect.

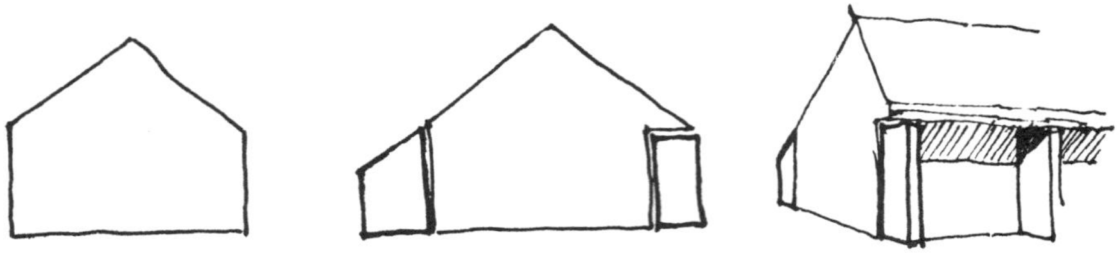

Wing walls break up long flat walls, make nice shadows; cut wind, and increase privacy. There is, however, some loss of sunlight.*

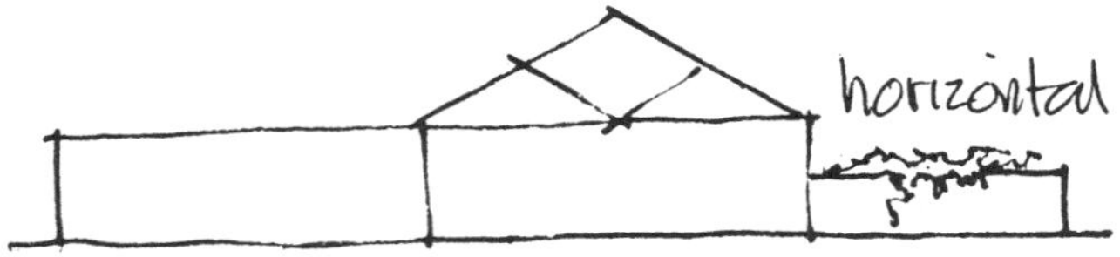

horizontal wing walls!

"same angle, cuts back," he says.

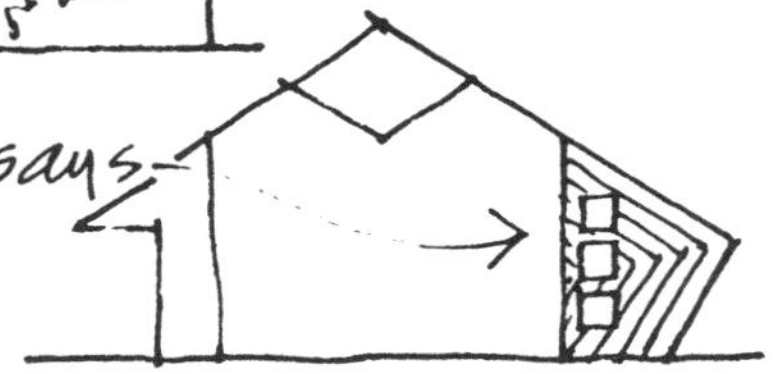

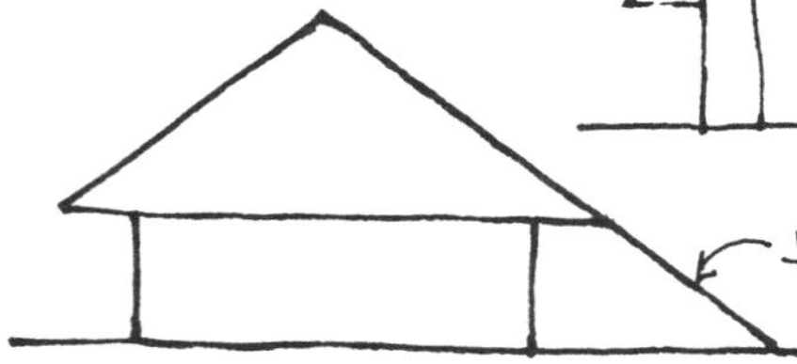

to the ground!

Wing walls can even be widened to create extra closets or outdoor storage spaces. The walls tend to hold the house more closely to the earth, visually. And here's an idea of Charles's to use wing walls to support a deck. He sure knows his wings, doesn't he?

You can cut holes in wing walls – or just carry down the beams. Open beams help carry deck and emphasize shade.

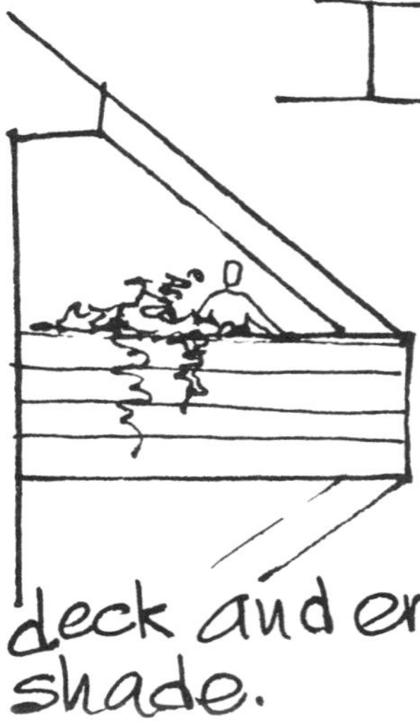

*As we've been saying, everything carries with it an environmental price tag.

South perspective of a Woodsian interlude!

Hawke Residence (project). Terraced residence.

"Watch those downspouts",* says Charles.

Well, I watch them all the time now and nothing much ever happens. But that Charles, he'll see things ordinary folks don't see, and it pays to listen to him.

He says to limit the number of gutters you use, and try to conceal them, or at least make sure they have more or less the same colors as the houses on which they're fastened.

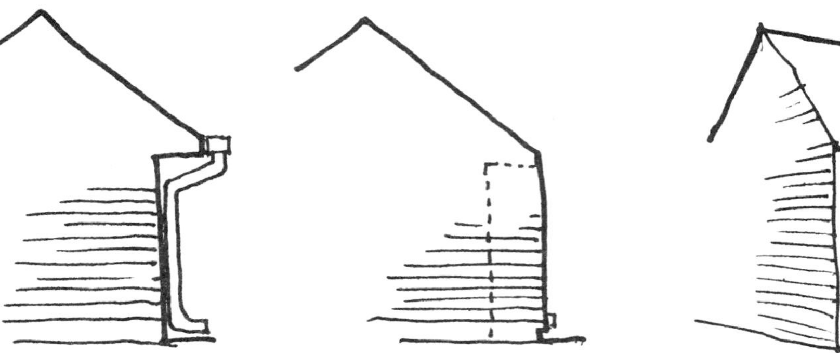

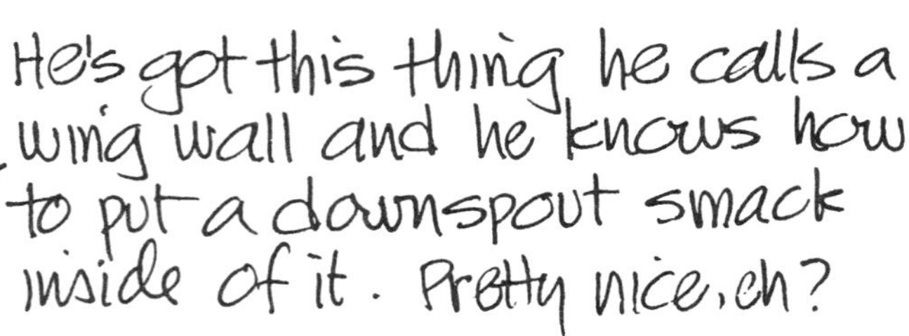

He's got this thing he calls a wing wall and he knows how to put a downspout smack inside of it. Pretty nice, eh?

* While you're watching those downspouts watch where the water goes, too. That's the main consideration.

Conceal downspouts in wing walls

THERE IS NO DOWNSPOUT ON THIS HOUSE

Charles Woods has a very discerning eye, an eye that is offended by the little distractions in a house design that many of us might never see. When he put this rule on the list I asked him why he'd bother to hide something as innocuous as a rainwater conductor. His answer took us through the dialectics of philosophy into a history of design, back through Goethe, Nietsche, Mann, and Engels, into the realms of metaphysics. Only then did I understand: he didn't like the looks of downspouts.

Well, I must say he has a point, there. The wing wall *does* look better than a strip of bent and corrugated aluminum tubing.

"Is it dishonest to conceal a gutter? Frank Lloyd Wright was very honest in his architecture but he concealed electric, plumbing, heating, etc., and said something like 'our intestines are supposed to be concealed.' I agree." —Charles.

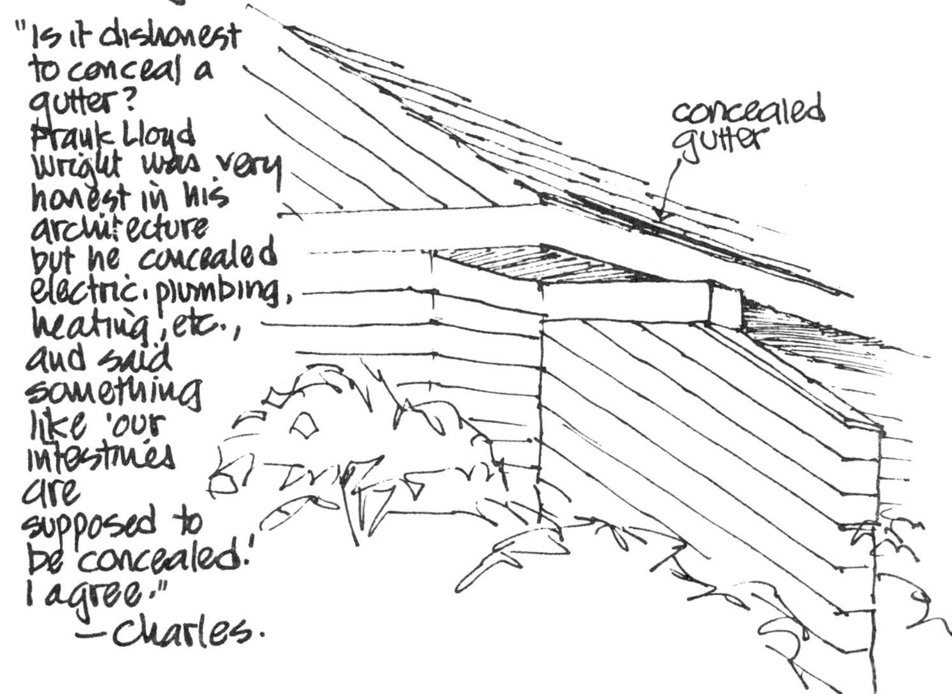
concealed gutter

Does the rainwater rush away, wasted, or does it soak into the soil as it should?

More on proportions.

Three walls, three different height-divisions: 1/2, 2/3, and 1/3. Is one better than the others? Yes, but who can prove it? Can you *feel* that the first one is indecisive, uninteresting? It doesn't know if its a brick wall or a wood wall. The other two leave no doubt; one is a brick wall with a bit of frame, the other is a wood wall with a brick base. They are not ambiguous. They are architectural.

You've seen mats around watercolors. There's usually a wider border at the bottom. It just feels right. Do the same with doors, or panels. And the more elongated they are the wider the top and bottom borders should be. Carry out the design theme.

Charles adds, "Follow Da Vinci's Golden Mean!" I say don't get lost.

From to to , all the lessons in proportion we'll ever need are out there waiting to be rediscovered by us, the lost tribes.

Make the garage match the house.

The garage, or any wing or accessory building, should always be of the same design as the house. When it's possible, the roof angles, the windows, and of course the colors should be those also used in the house. And remember: Charles said that the garage should not be larger or wider than the house. He added that on contemporary houses the garage doors can be flush or siding covered (if you add stronger lift-springs).

(Too big.)

Let's see if we can find an example. Ah, here we are...

Insulate garage doors if garage is heated

And here's another.

(Le garage.)

Granted, this is not low-income housing but let's just take a look, OK?

Cars are made to be left out of doors. Do you <u>need</u> a garage? Think of the materials, the cost, the junk you'll collect, in a garage.

Corner shame must end!...

Why do we cover all the corners and edges with boards and mouldings? Are we ashamed to see the great mass of the building turn the corner before we get into the eaves, gutters, etc.? Boldness, as opposed to fussiness, is what gives old barns their charm. We can have it in houses, too.

...as must stair pollution.

← Aren't those pointy arrogant stair corners a bit hostile?

Why not let the plants cover those corners? Why not try to express, in our landscape features, a respect for the world we trample so casually with our architecture?

← Charles prefers 1½" round steel pipe railings.

Here in the temperate northeast it's fascinating to watch the succession of plants when a lawn is left unmowed. Grass this year, then wildflowers, then trees...

Posts, columns, supports.

Tell the truth. If it's a wooden post, let it be a wooden post. Don't wallpaper it with thin pieces of stone, or horizontal or diagonal boards.

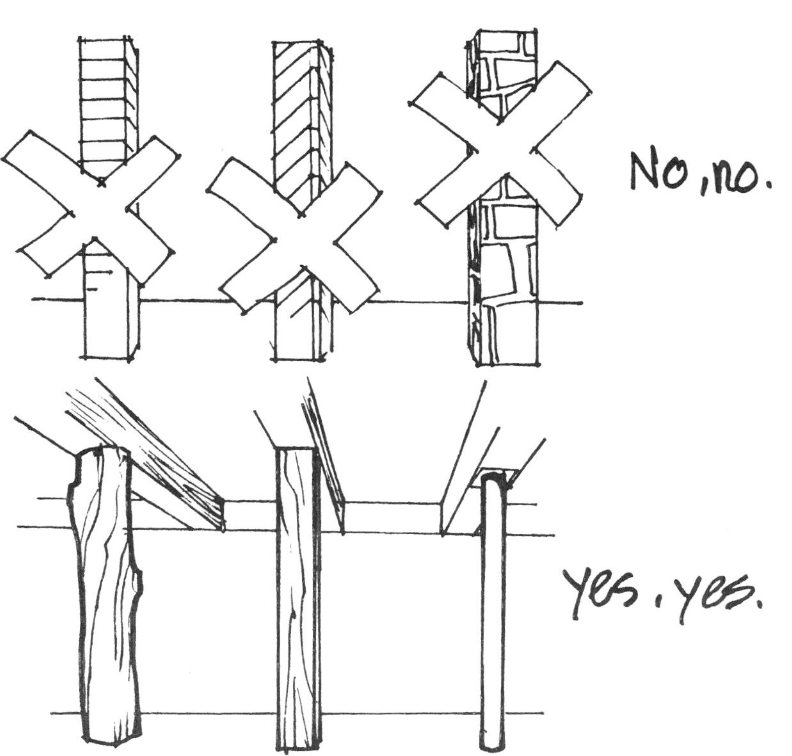

Many of us live where stone is available just for the taking. If that's the case where you live, why not take it? There's no involvement in your house project as satisfying as building part of it yourself. Stonework is so elemental, so basic and rewarding, you really should try it. Only a few basic rules need be followed, and the results will amaze you. It's not hard to see which wall is ready to fall apart, is it?

"I think stone columns should be at least 12" wide." —CGW

Stone is a low-environmental-impact material. And if you have time to be your own mason you'll find it a highly rewarding experience.

Use gentle stair proportions.

The first rule of steps is: are you creating a barrier... for the handicapped, for furniture moving? Is their another way to get up and down?

more like these:

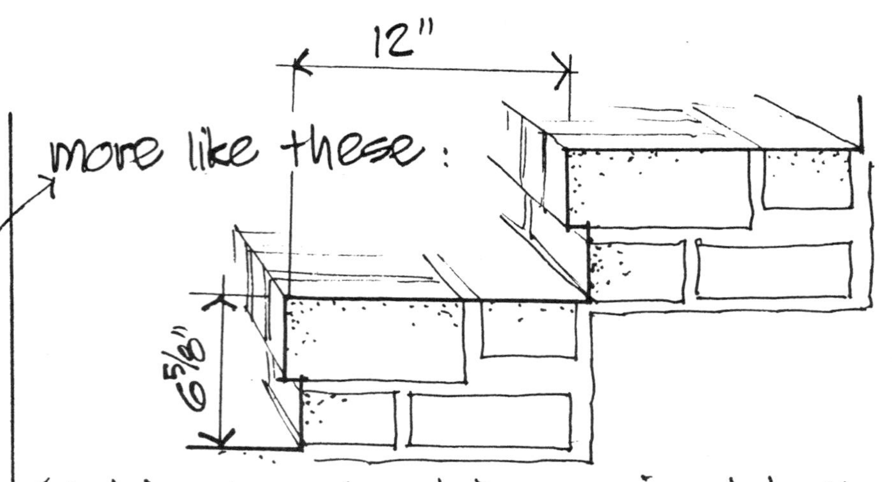

(Notice that the dimensions in both cases ignore the overhangs.)

No? OK. Then build stairs that are both safe and comfortable to use. Indoors, that means riser and tread proportions close to this.

10½" (9" min.)

7½" (8" max.)

Outdoor steps are generally less steep, with proportions

Those steep old stairs may be quaint but they're no fun, and can be dangerous.

How did they move their furniture up and down?

By learning to live on two or more levels we did, in a sense, halve our land use. Population reduction would be a better way.

woodsian interlude.

"H - Wedge."

Use great care in locating the house on the site.

suppose the lot you buy is in a neighborhood of houses

that uniformly face the street but you want to take advantage of solar energy by facing a wall of glass toward the south. What do you do?

How about this?

low roof with sheltering earth berm on this (cold) side.

Location - siting - involves more than just where on the lot you set the house. There are vertical considerations, too, and not just the obvious ones of groundwater depth and proper drainage. You want the house to appear comfortable on the land.

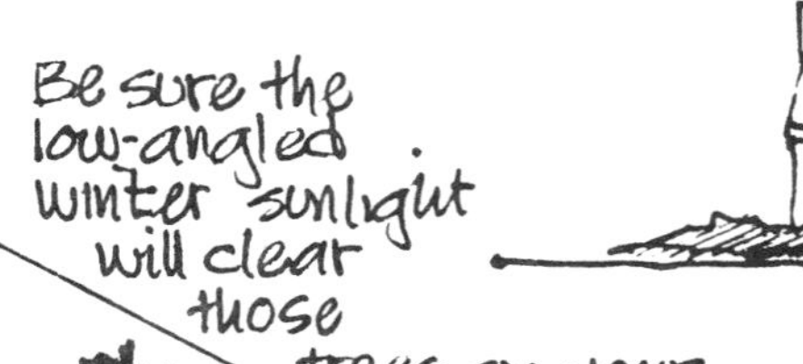

Be sure the low-angled winter sunlight will clear those trees on your neighbor's lot.

Saving energy is reason enough to use common sense in site planning. The bonuses are better appearance and perhaps healthier land.

Keep the street side of the house private.

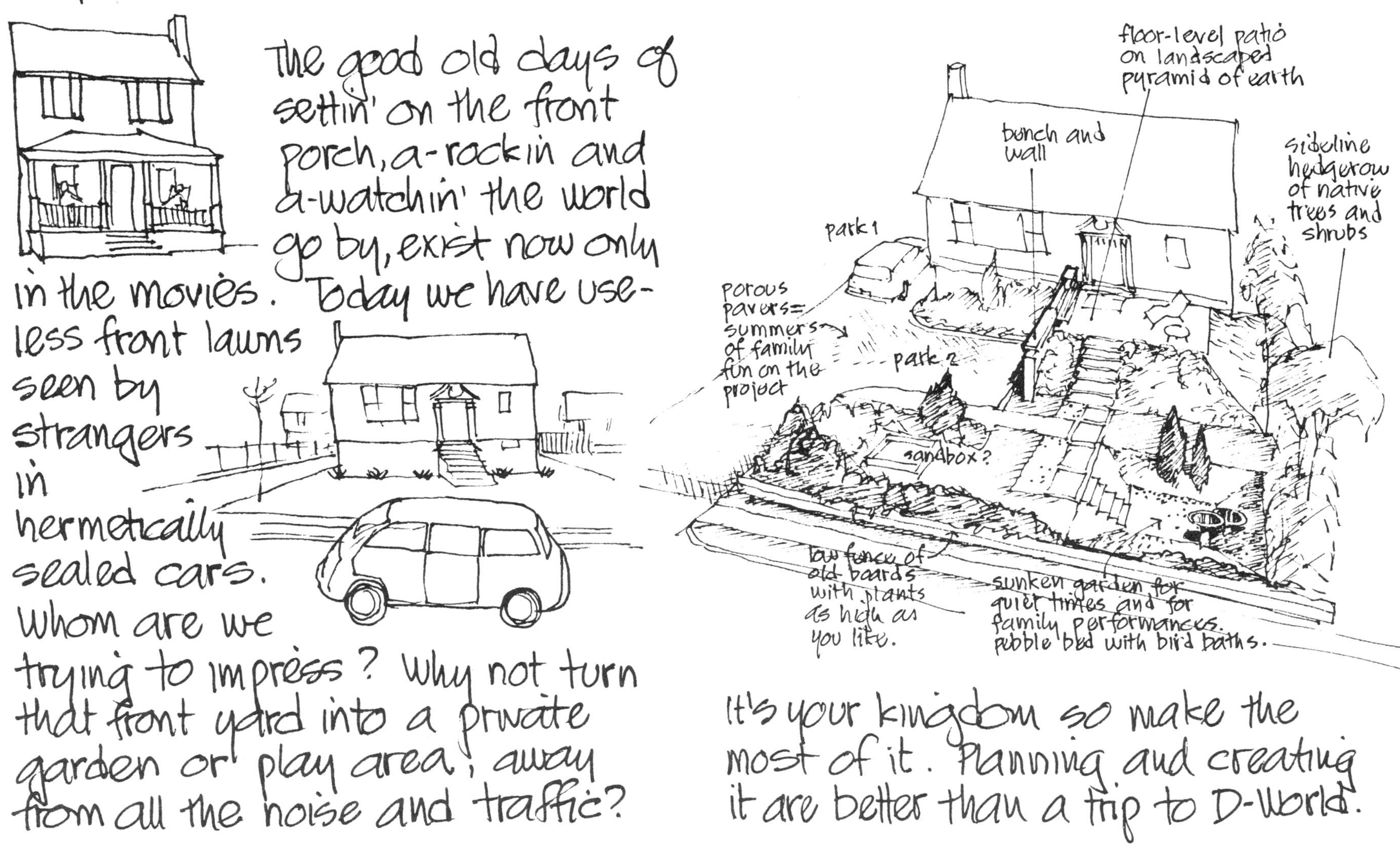

The good old days of settin' on the front porch, a-rockin and a-watchin' the world go by, exist now only in the movies. Today we have useless front lawns seen by strangers in hermetically sealed cars. Whom are we trying to impress? Why not turn that front yard into a private garden or play area, away from all the noise and traffic?

It's your kingdom so make the most of it. Planning and creating it are better than a trip to D-World.

You'll be amazed and delighted to see all the wildlife that will move back onto your property once you put the mower aside.

Backyards

Charles asked me to show a private backyard, too, with the house set as far forward as the zoning will allow.

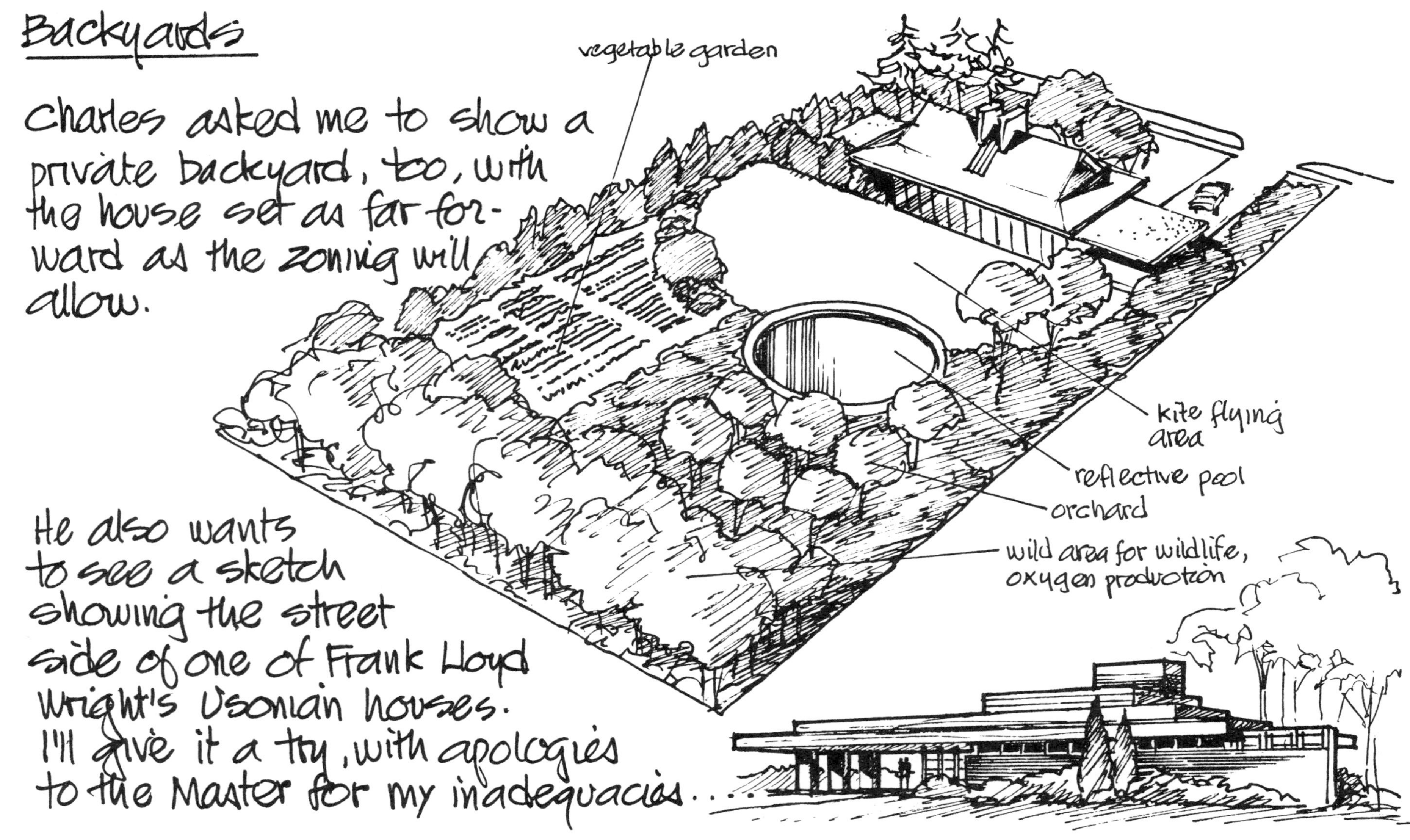

He also wants to see a sketch showing the street side of one of Frank Lloyd Wright's Usonian houses. I'll give it a try, with apologies to the Master for my inadequacies . . .

How is all this being received in East Los Angeles or the South Bronx . . . or by the homeless? Are we being totally irrelevant?

soften those rectangles and straight lines!

Happy, easy, do-it-yourself weekend projects as easy as a, b, c. The observant reader will at once detect that I have reversed the drawings on the bottom line. - M.W.
a.
b.
c.
shelter!
grapes!
c.
b.
a.
Ever make a huge earth-cast concrete bowl?
All you need are a couple of bags of sakcrete.

An above-ground house can look nice but it can be a maintenance nightmare. Arrange it so it looks poetic but leave accessways.

Septic mounds: design and place them carefully.

Know what a septic mound is?

No? Well, then you needn't even bother with this page. Lucky you. Oh, well, OK, I'll explain: people who continue to drop their body wastes into pure drinking water instead of into silent, odorless composting chambers must upgrade their septic tanks and disposal fields as the old ones clog, leak, or break down. If the soil is impermeable or if the groundwater level is too high the disposal field must be made of imported fill piled several feet deep over a large area of the property, advertising one's shameful practices. Here's how they often

look:

...and how they can look:

"I put a Japanese zen sand/rock garden on mine", said Charles.

If you have city sewer service don't gloat. It just moves the wastes somewhere else. How many "somewhere elses" are there?

Driveway improvements,

If your driveway is too steep, or too long, or too impervious to rainwater it just isn't right.

too long and boring

too steep

too impervious

An asphalt tide is spreading across America. Powerful interests will see that it continues to spread. The best you can do until the national mood changes is to make your paving useful, interesting, and

"Mac, highlight this ★*★ Important! Almost everyone does this."

"Also," says CGW, "use red shale instead of modified. A sea of gray is ugly. Use modified as a great base."

less destructive.

Modified what?

You can make a drive less steep by lowering the garage, perhaps, or by taking a longer, curving approach. Such an approach would do wonders for the long, boring drive at left. And porous paving blocks allow grass to grow and rain to penetrate the surface. Set the blocks on crushed stone, add topsoil and seed.

– Does what? "Driveways too long, too straight, too boring."

Porous paving blocks, variously known as Grasstone, Turfstone, Turfcrete, etc., are available from most concrete block manufacturers.

Keep the landscaping simple...and appropriate!

Let's not force plants to commit unnatural acts. No more maple trees in the Arizona desert, OK? No more Colorado Blue spruce in Boston. And no Japanese red maples except in Japan.

If you follow the basic rules you can't go wrong.

1) Use trees and shrubs that are native to your area. They'll do well there because they evolved under local conditions.
2) Don't try to exhibit a plant of every species in the book. Let the local nursery do that.
3) Plants of all kinds should be arranged mostly in clumps of their own kind, the clumps blending into clumps of other kinds. Look long and hard at the most appealing of natural (wild) landscapes. They're never just mixtures of plants. They tend to have areas of one species edging into areas of others.
4) Never, never do an "engineered" landscape, with trees or shrubs spaced dot-dot-dot in straight lines. Landscaping, to be appealing, must appear to be almost a random arrangement. The observant visitor will know that the randomness is in fact a controlled design, and a very successful one.

Native plants not only look appropriate, they act appropriate, too, needing little, if any watering, and no pesticides or chemfertilizers.

This is bad.

And this is bad.

So is this.

You wouldn't think of treating a friend that way so why do it to a beautiful plant? OK, so we've shown some flat-top hedges in these pages. Well, um, maybe when you really need an architectural line in the land-scape a trimmed hedge is forgivable, but don't trim everything. Let Nature show you what she can do.

And for heaven's sake, throw that land-scalper away. You can mow the bit of lawn you need by hand.

Remember: these green things are not building materials or decorations. They are fellow earth-creatures on which our lives depend.

Score lines in your dyed concrete floor.

"How do you do that?," I asked, and Charles was ready for me: "Cut or trowel them in."

There is a cement finishing tool - a type of trowel - that will put a slightly rounded edge on such grooves.

On the other hand, cutting the grooves, using a carborundum blade, after the slab has set (hardened) shouldn't be too bad a job if you're wearing both a mask and goggles.

I wonder, though, how you manage to cut the groove all the way to the wall. Doesn't it stop about 3" from the vertical surface? I'll ask Charles.

"Pour the floor slab first, and cut it. Then build the walls, a la FLLW."

Aha.

Every hard surface in a room adds to its acoustical problems. Clatter! Books, upholstery, drapes, rugs, and wall hangings help offset this.

From Charles:

"Gray concrete walks are ugly...

"...dye them," says he, "or expose the aggregate (the stones and pebbles in the surface of the mix)!" Round brown pebbles are his choice for exposure. Most cement finishers know the technique of exposing those pebbles without loosening them, sometimes ordering a stony mix, sometimes adding the stones after the slab has been poured.

I've found dyes to be unpredictable so I stain the concrete after it's been cured. A watered-down stain made of ordinary latex house paint in one or several earthy tones is appealing and it lasts many years.

Gray concrete often strikes us as ugly because we associate it with ugly places but if you do it like this the gray will be forgotten:

If my warped drawing above↗ has made you dizzy please sit down for a moment and eat an organic carrot.

"Gray concrete" (cont.)

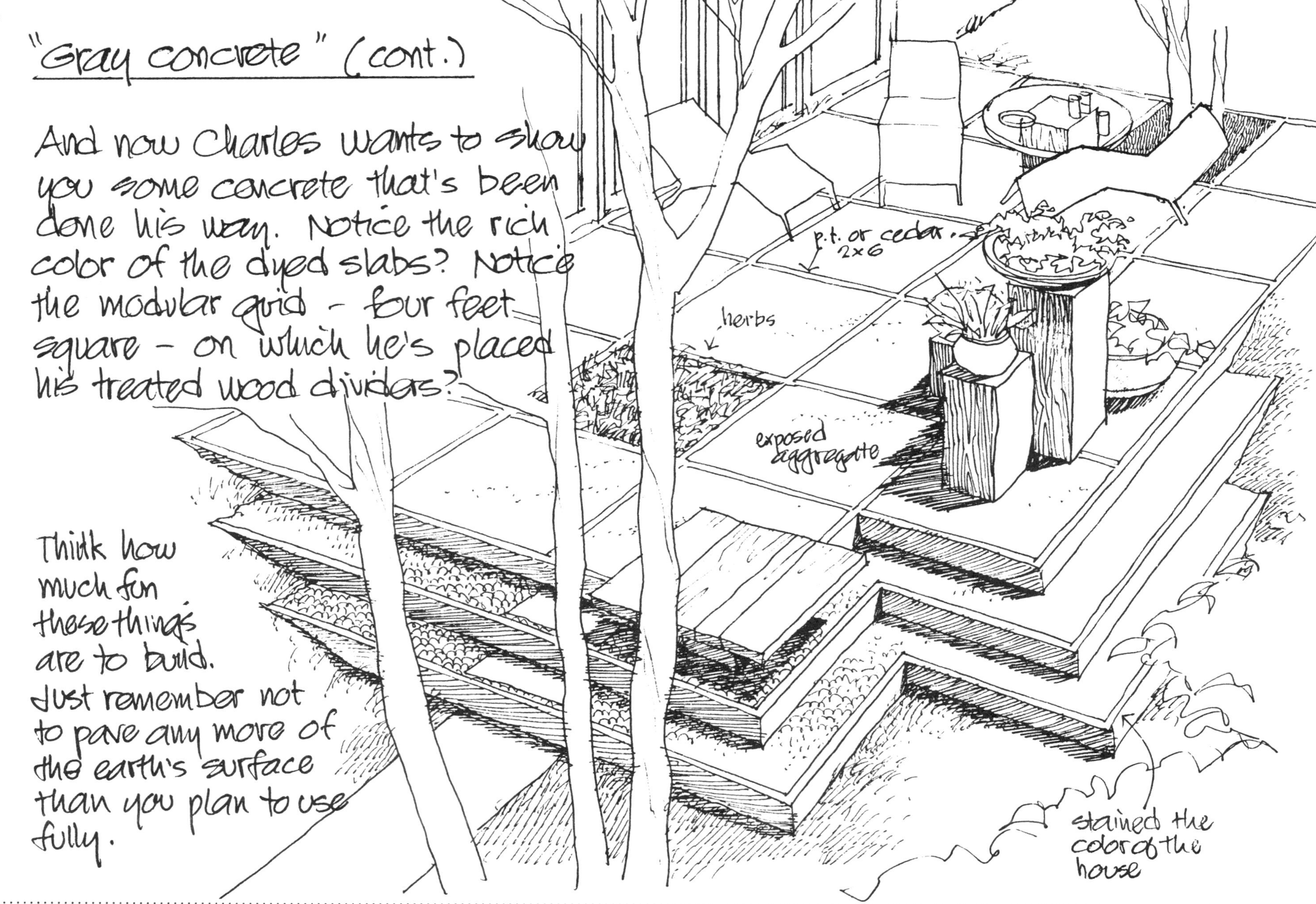

And now Charles wants to show you some concrete that's been done his way. Notice the rich color of the dyed slabs? Notice the modular grid – four feet square – on which he's placed his treated wood dividers?

Think how much fun these things are to build. Just remember not to pave any more of the earth's surface than you plan to use fully.

A sensitive and gentle design will get so good, over the years, as it grows into the landscape, it will seem to have been there forever.

A mild disagreement on the subject of masonry.

Charles and I agree, right down the line, on the subject of stonework. The first rule is never to use that fake stuff that's really cement stucco shaped and painted to look like stone. You've seen it:

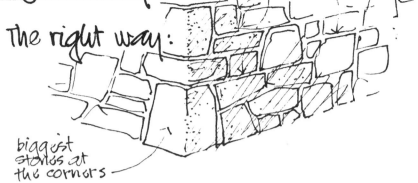

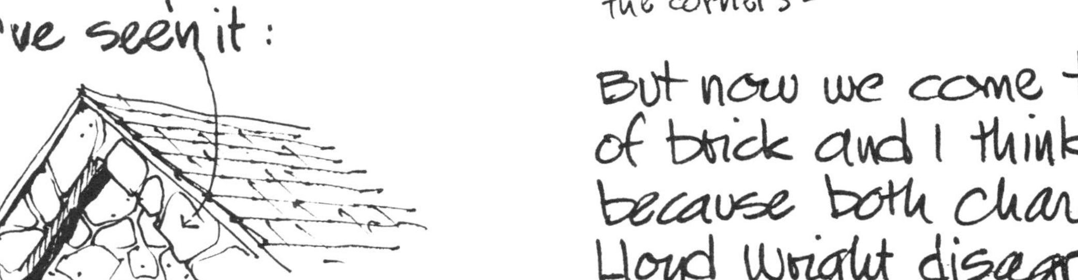

The right way:

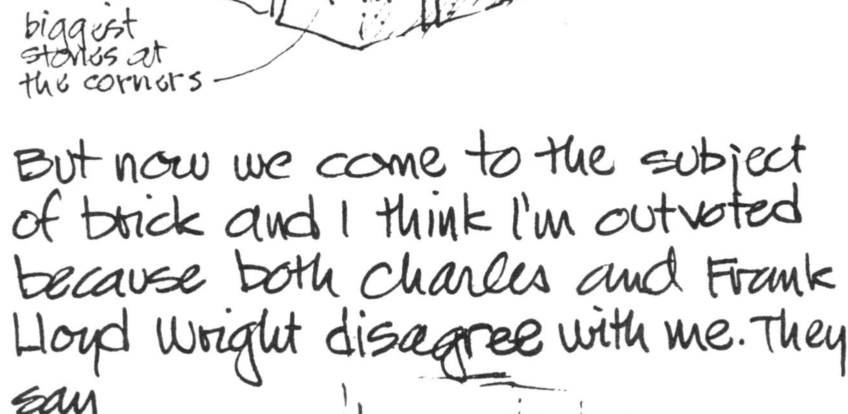

Rule 2: Lay stone flat, never like this. It looks like potato chips this way and is very unhappy:

But now we come to the subject of brick and I think I'm outvoted because both Charles and Frank Lloyd Wright disagree with me. They say you should rake out the horizontal joints.

I like brick with all its joints cut off flush, so the wall itself is one great masonry surface.

Charles says you should dye the mortar the color of the brick or stone, or slightly lighter.

Insulation outside the masonry is much more efficient but it conceals a building's character. It's a new challenge for designers.

Hug the ground.

Establish floor elevations carefully. Basements should not stick up out of the ground. Houses are part of the living earth. Let them look that way.

Boo!

Awful. Look at that spindly-looking deck support!

You can relate the house to its site either by lowering the building or by raising the grade around it. Your decision should be based in part on the presence of groundwater. If a basement floor is below the water table it is due for a flooding sooner or later. Pumps and waterproofings aren't perfect.

See how much better even that silly house looks when properly related to its site? And it has a good start toward successful landscaping. See how it all begins to come together?

Now to fix some other mistakes...

The less exposure a building has to winter winds and summer heat the more energy it will save. Earth is a good protector.

An open stair to the basement is nice... if you have a basement.

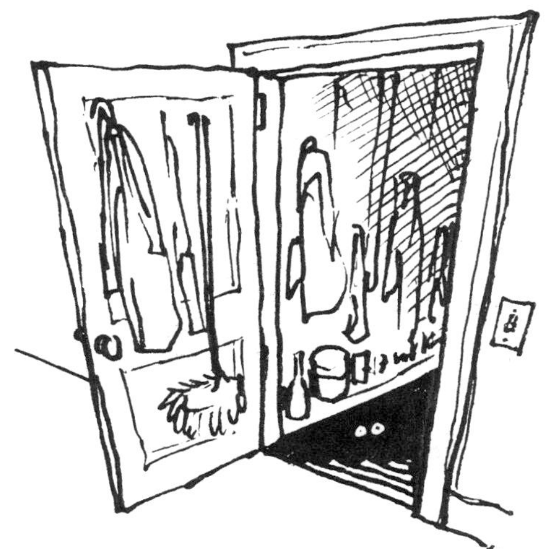

The stairs to most basements are enough to scare little kids to death. Dark, cluttered, and dangerous, they give basements a bad image unnecessarily; basements are usually bad enough to create their own gloomy reputations.

Open stairs, on the other hand, ...

...make basements extensions of the main floors.

Charles reminds us that you can have skylights above these stairs, planters in their walls, bookshelves - even windows on the landings.

Down into the loving arms of the earth — beneath the sunlit land. What promise the prospect holds!

Let 7 feet be the minimum height in your basement.

You can avoid the headroom restriction imposed by a center beam by using joist hangers

And Charles adds this: don't exceed 8 feet unless necessary. (I always recommend 10 to 12 feet where elephants are involved.)

but then all the pipes and ducts must go under the beam. Better to let them cross over:

"Too tall a basement makes the house look like a saturn rocket," opines Charles. "If you need a high space in the basement step the floor down."
Charles says move the window down and put up a drywall ceiling.

Basements are huge, 3rd-class spaces under first-class spaces. Isn't there a more efficient way to use our resources?

Consider slab-on-grade rather than wood frame floor construction.

OK, you've considered it long enough. Which one have you chosen?

This? Or this?

The cheapest, and, in a survey, the most appealing, floor is the polished leathery-brown finish of mixed wood-stain and urethane sealer on rough, not troweled, concrete.
high-R insul. bd.
ground level
termite shield
crushed stone and vapor barrier
rigid insulation
vapor barrier must be on warm side of insulation.
basement
(or crawl space?)
vapor barrier, plus dampproofing in well-drained soils, or membrane waterproofing where ground water is present.
reinforce wall as req'd to meet load conditions.

Don't forget to provide convenient space for the segregation and storage of household wastes to be recycled.

More about slab-on-grade vs. basements.

Here in the northeast, where Charles and I do most of our work, foundations must go 3 or even 4 feet into the ground in order to get below the depth at which frost could go and heave them.

Basement advocates argue that if you must go that deep you might as well go a few more feet and have the advantage of all that basement space. It's a good argument, as far as it goes. But Charles and I are about to let you in on a secret process that will demolish that argument:

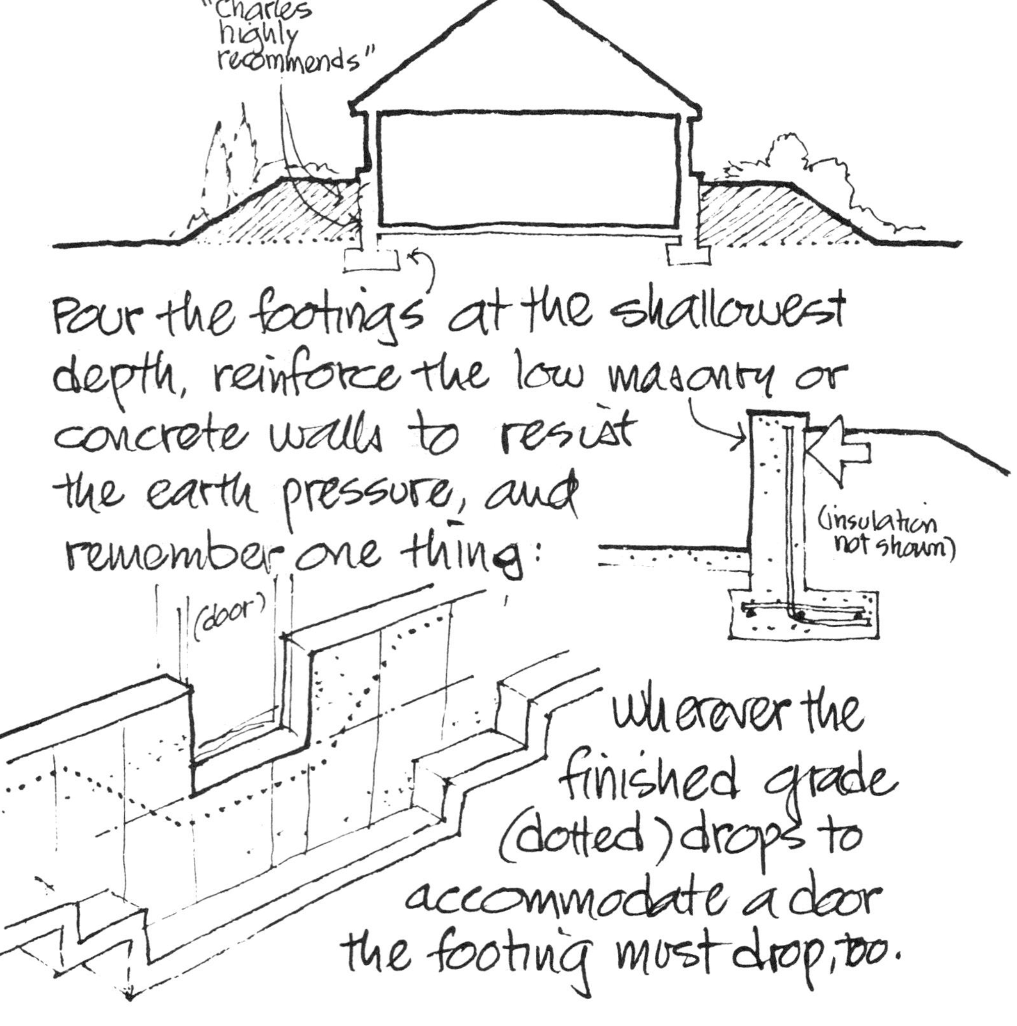

Pour the footings at the shallowest depth, reinforce the low masonry or concrete walls to resist the earth pressure, and remember one thing:

Wherever the finished grade (dotted) drops to accommodate a door the footing must drop, too.

The earth's population is due to double in less than 40 years. Should we be doing something about it now?

More on basements.

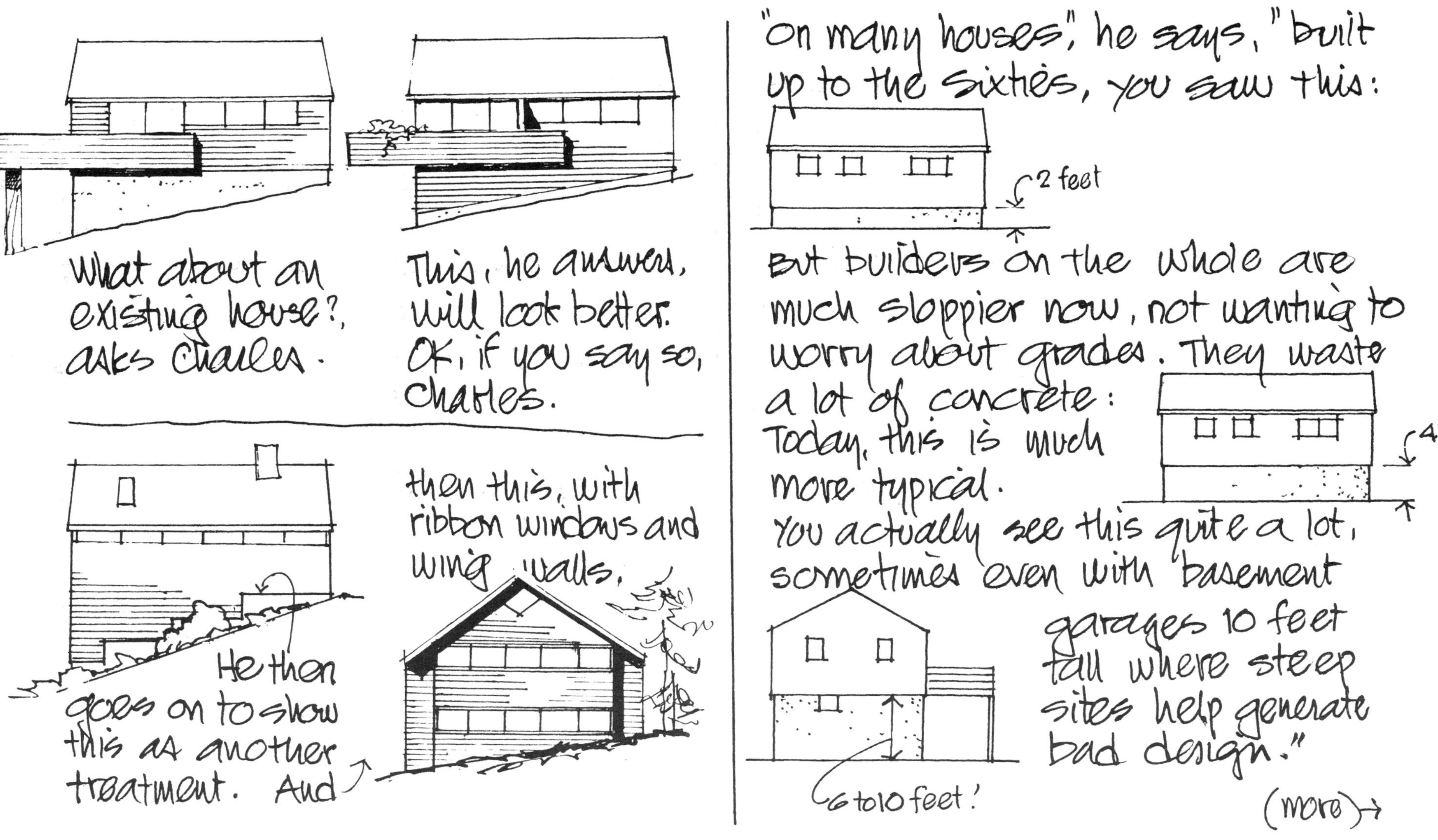

What about an existing house?, asks Charles.
This, he answers, will look better. OK, if you say so, Charles.
He then goes on to show this as another treatment. And
then this, with ribbon windows and wing walls.
"On many houses," he says, "built up to the sixties, you saw this:
2 feet
But builders on the whole are much sloppier now, not wanting to worry about grades. They waste a lot of concrete:
Today, this is much more typical.
4'
You actually see this quite a lot, sometimes even with basement garages 10 feet tall where steep sites help generate bad design."
6 to 10 feet!
(more)→

You almost want to give up when you see how much worse residential design gets each year but if it were getting better who'd buy our book?

"Why call it a basement?! And why build it out of concrete if it sticks that far out of the ground? It's not only more expensive and harder to cover with siding, it creates more of a moisture problem, too."

Charles has designed for us this alternative to the house shown above:

Quite a difference when Charles goes to work on it, isn't there?

A properly designed (and built) basement will not leak, and will, as a result, be easier to heat, and less troublesome in the summer.

No plastic house columns!!!

Now what does he mean by that? (I'll have to recheck my notes.) What else would a plastic house use for support but columns?

I can see it now: a snap-together house made of bright extrusions.

Charles? What do you have against the use of columns on a plastic house, honestly?

"Nothing", says Charles with a straight face. Then he explains that he has actually seen a building with classical columns made entirely of – would you believe it? – plastic! "Huge ones," he says, "more than 12" in diameter and 16 feet tall! Like the one at left."

Charles goes on to remind us that ornate iron columns are out there too, waiting to seduce the unwary, so remember now that you were warned, right here, against ever using such things.

From Portland to Portland a sea of look-alike cracked plastic is starting to cover America. Can we hope to slow it? Stop it?

What do fake plastic-foam wood beams say about us?

What about plastic "rocks", "slate" made of rubber, and asphalt shingles grained to look like wood? Well, to be consistent

we should carry them home in a station wagon (or van) that has fake wood grain on its sides. That way, people will know that we are, at least, consistent. They'll know that our houses are probably full of materials and furniture, clothing and appliances, that are non-natural too, giving off all kinds of fumes, toxic materials, radiation, and particles that are all part of the good life. The good plastic life, that is. It's a period we must go through before we mature as a nation and learn to live in healthful houses and to care about the natural world from which we think we have escaped.

For a good book on non-toxic houses, read David Pearson's THE NATURAL HOUSE BOOK. Simon & Schuster. 1989. 287 pp. $19.95 ISBN: 0-671-66635-5

Watch out for fancy windows!

The round-top fad of the 1980s is already starting to mock us. What in the world will be next, star-shapes, or triangles? Where do we get these bizarre ideas? Are our houses not gaudy enough?

I'll draw a few here that were suggested by Charles and see if he approves. If he doesn't you'll see an obvious patch where I tried to glue a revision over the original.

An opening of any shape can become an appropriate part of a design as long as it is clearly a part and not an unneeded feature.

But you get the idea: Unity.

And then, if you can make everything blend into a native landscape you'll really have something.

Watch out for fancy windows (cont.)

"Mac, I agree with you on this but think interesting windows can be nice *if* they're integrated into the house design. Show some of these and say 'Charles likes'"?

What do you mean you agree with me? *You* wrote the other page. Anyway, OK, here we go. Charles likes:

skylight beyond

Instead of this:

or this:

although this is better!

Charles! Your vines are growing down into the environmental area. Can't you spray some herbicide on them?

Protect exterior doors with overhangs.

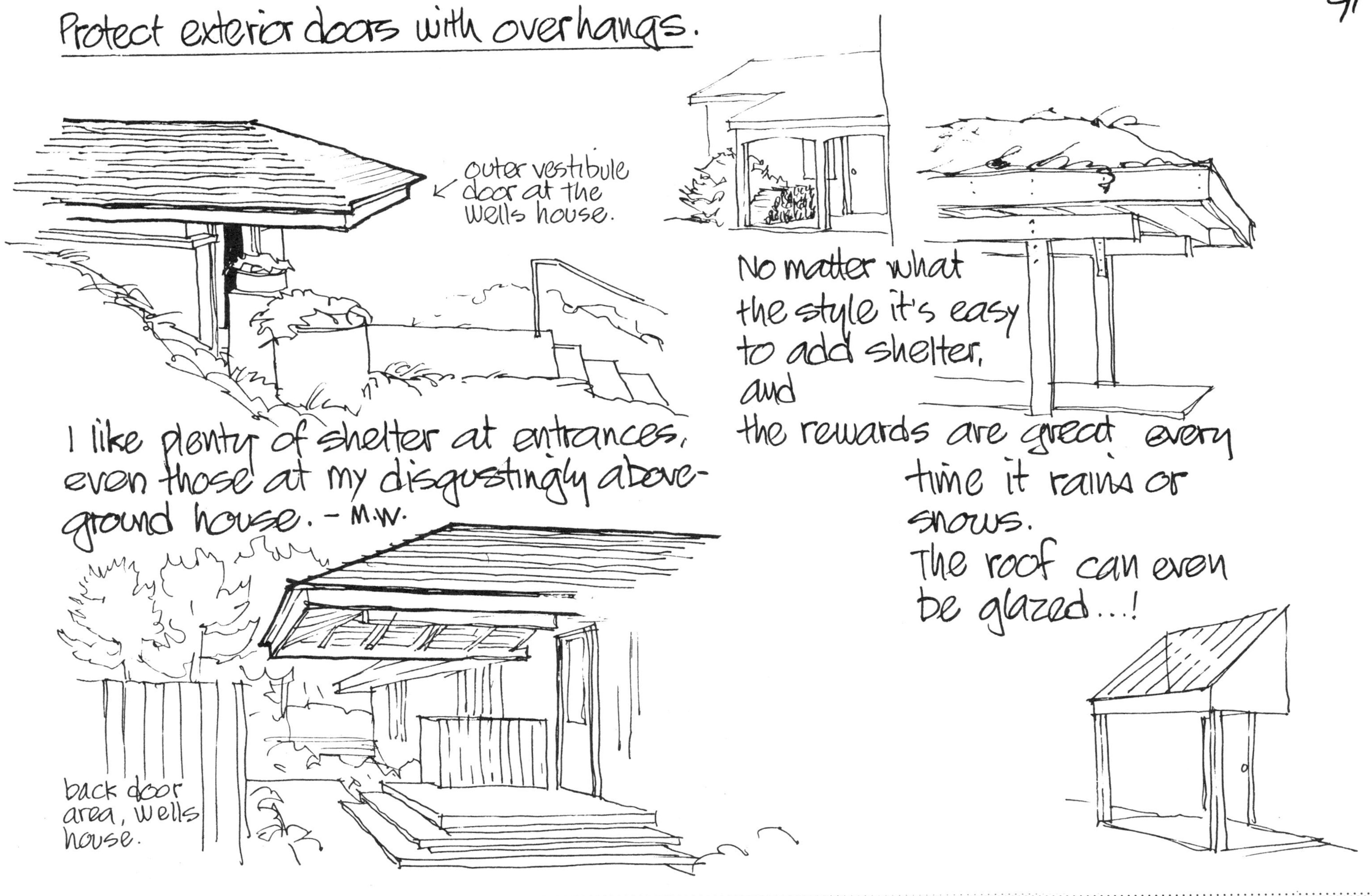

I like plenty of shelter at entrances, even those at my disgustingly above-ground house. – M.W.

No matter what the style it's easy to add shelter, and the rewards are great every time it rains or snows. The roof can even be glazed...!

There is one profoundly simple reason why I am again temporarily living above ground. It looks like this: $. Lack of it, that is. But my office is under.

Use an airlock vestibule for wintertime energy efficiency and comfort.

Without such a vestibule, a house can be swept full of frigid air in just a few seconds when a door is left open.

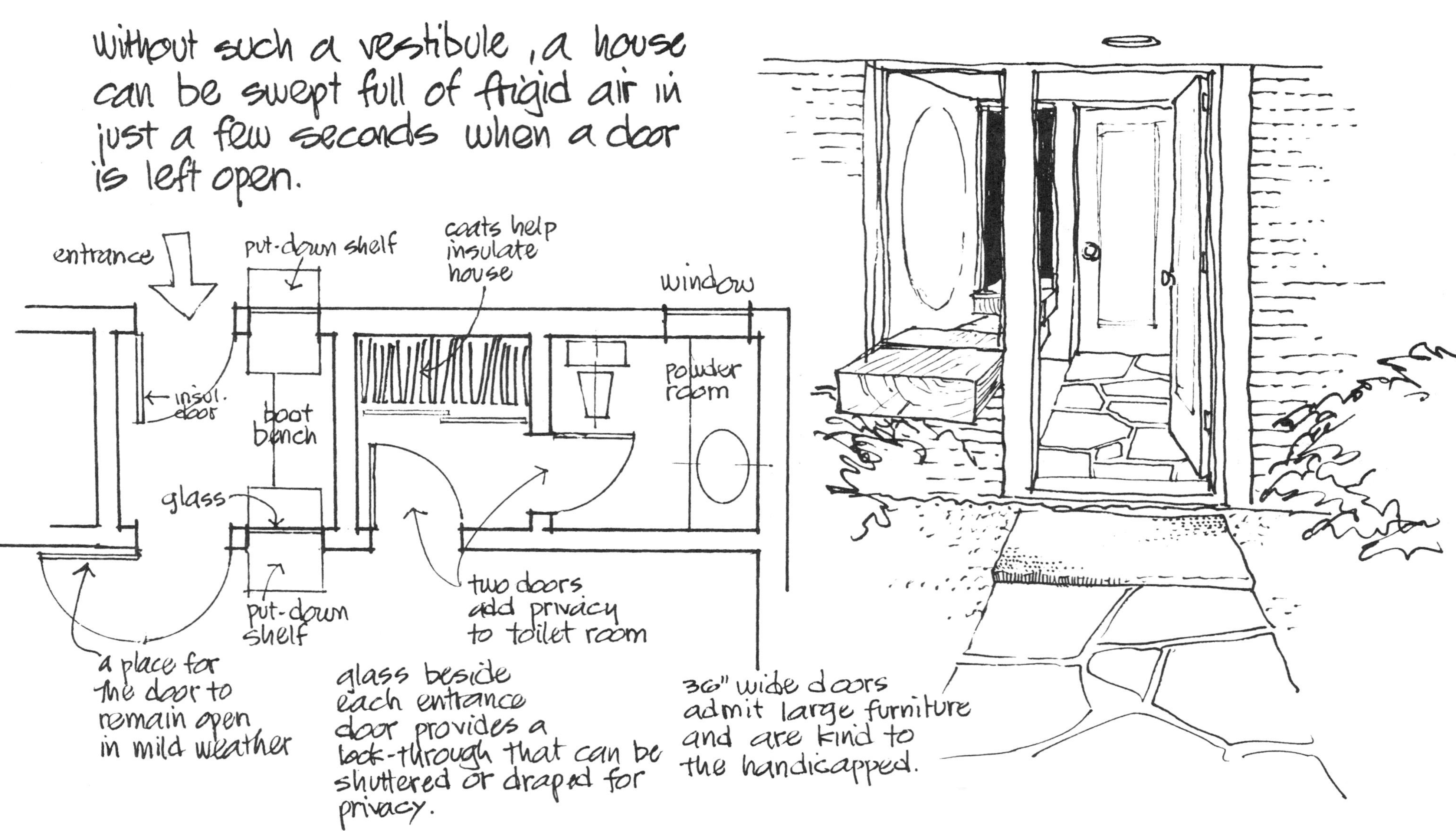

entrance
put-down shelf
coats help insulate house
window
insul. door
boot bench
powder room
glass
put-down shelf
two doors add privacy to toilet room
a place for the door to remain open in mild weather
glass beside each entrance door provides a look-through that can be shuttered or draped for privacy.
36" wide doors admit large furniture and are kind to the handicapped.

Somewhere on earth an oil spill didn't happen, and the air got a little cleaner because you remembered to build a vestibule.

Use good quality doors and hardware.

Charles and I say they're worth the money, especially for exterior doors. Interior doors should be heavy and solid*, and well-fitted, too, especially if sound transferance could be a problem. But for many interior doors, shed doors, etc., all you need are a couple of hinges and a turn-block.

doors should have 4 parts: the door frame, the door, the hardware, and the <u>put-down shelves.</u> Without that 4th part you'll always have to put your armload of stuff on the floor while you fumble for the key – or the doorknob.

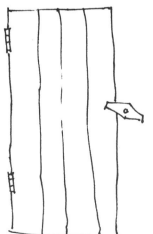

Just remember that ↗

*"Or," says Charles, "<u>good</u> hollow core with cardboard fill."

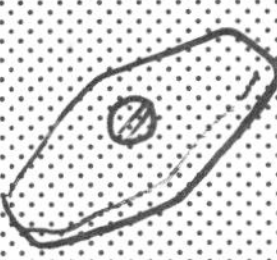

The old wooden turn-block will probably outlast the high-quality hardware, be easier to repair, and cause less environmental damage.

Don't use fake shutters.

The Indians may attack at any time, and then where will you be?

If you feel you must decorate your windows, make sure they're this type and not casements, awning, or picture windows. Doing this, for instance, makes the whole world laugh at you:

double hung

so does this:

If you use shutters, they will look appropriate only if they are made of wood (not plastic or metal), are of the right size (half the window width), and fitted with shutter hardware (which you can buy from the village blacksmith under the spreading chestnut tree in E. Laptop, Connecticut).

Indoor shutters... tight-fitting ones that insulate... can slash fuel bills and make a house much more comfortable in cold weather.

"Use flat glass in roof windows and skylights," says Charles.

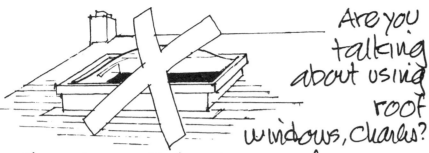

Are you talking about using roof windows, Charles?

They have flat glass, and you can open them, too. Still you can't beat those plastic bubbles for strength.

What bothers me about both skylights and roof windows is their placement. Here on Olde Cape Codde some people go to great lengths to build authentic-looking cottages, and then drop in a roof window that makes a joke of the whole design.

Make roof windows a part of the whole design instead by relating them strongly to all the rest.

Charles says if you do use roof windows on traditional houses keep them small. The windows, not the houses.

The Great Escape: wintertime heat-loss through rooftop glass. It's a real loser unless all known preventive measures are employed.

To bay or not to bay.

Many people who like modern houses like bay windows as well, so they buy standard bay units with angled sides that have no relation

to the rest of the house. It's too bad because it doesn't have to bay – I mean be – that way.

A square, not angled, bay window will tend to fit most modern designs and offer a choice of several interior arrangements.

They now make double bent corner glass

And there are 403 choices beyond these.

The winter night heat loss can be severe from a big glass box. Triple glazing, heavy full-length drapes, and window quilts help a lot.

Use good proportions

"This is in part subjective," says Charles, "but examples will help."

"a little short," he says. "Good" "a little long"

He's talking about floor plans, elevations, and windows. So let's ask him about the fact that most windows have shapes like this:

"Combine 'em," says Charles. "Don't be a hole-puncher. Let the lines of the house suggest the shapes of the openings. If it's long and low, let the window bands echo the design."

Here's the little house on Cape Cod that everyone loves. Its upstairs windows must be right on the floor! Is it nice simply because it's old? or did its builder have an eye for proportion?

Look what you see everywhere today.

It has a meanness about it that can't be put into words. Not here, anyway.

This is more like it.

Do good window proportions do anything for the natural world? Probably not, but if you install a composting toilet you'll do wonders.

windows should be continuous, centered, or patterned...

Charles, will you please come over here and show us exactly what you mean?

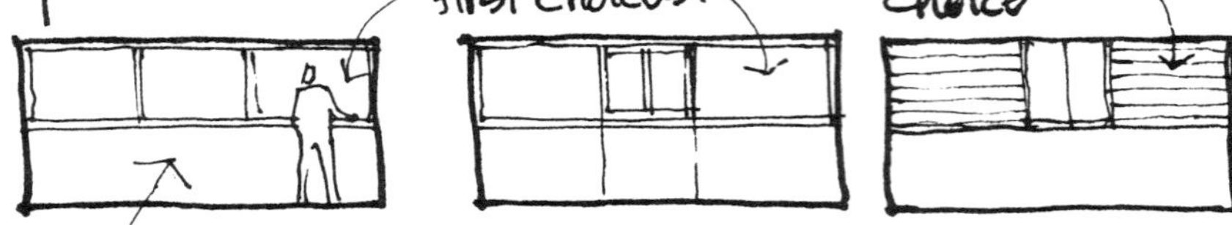

"There. That's what I mean. And to show you what I don't like, take a look at these two walls:"

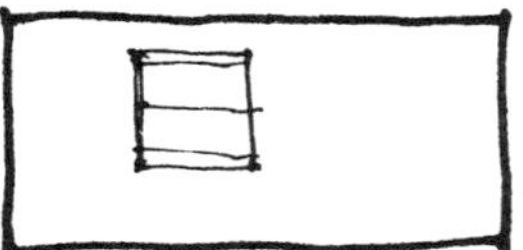

Thank you, Charles. I see that the two "bad" examples have no design, no order, no style, no relationship to the structure, dimensions, or lines of the house.

Now let's go outside and see if I can spot a house that illustrates what you don't like. Yes. Here's one: In fact, there seems to be a lot of them out here. A majority, in fact.

Now let's see some well-placed windows:

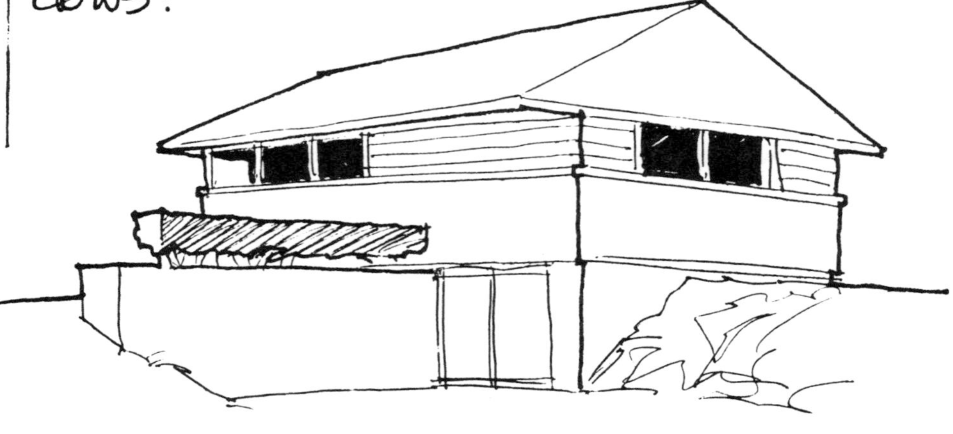

Does an orderly window arrangement do anything for the natural world? Probably not, but if you throw your lawnmower away it will help.

Don't mix window types.

It only proves you don't know what you're doing if you do. Try to make the house speak with one voice. If it says, "I am amateurish and chaotic" it's not saying the right thing. Restrain yourself when you visit the window department of the building materials center.

Charles says casement and awning wdws. combine well with fixed glass, double hung with fixed.

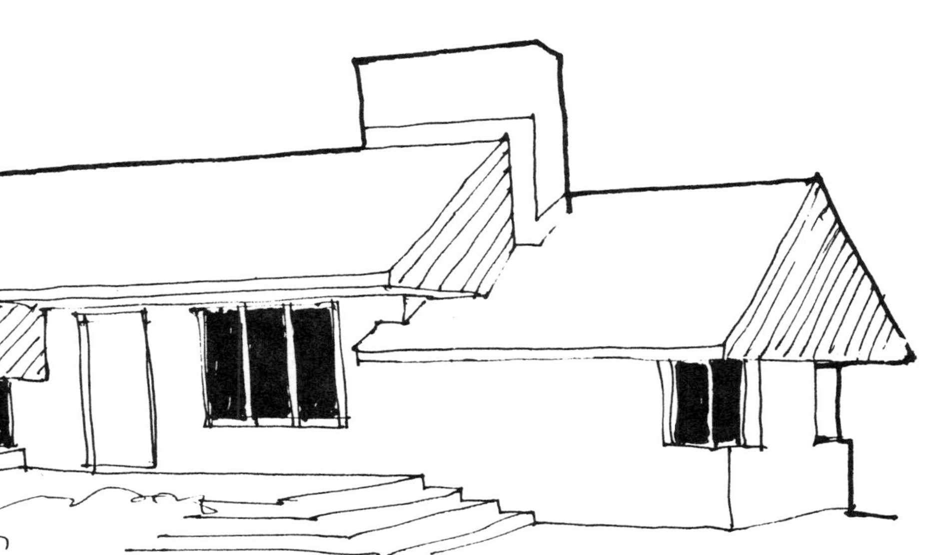

Here's a deliberately complicated little design, shown here to illustrate the way window unity more than anything, holds it together.

Windows are the biggest heat-leakers in a house. Shop for efficiency and think of weather extremes when you size and locate the openings.

Better basement windows.

Just because they're down in the cellar is no reason to use cheap windows. To be efficient, to last a long time, and to look good they must be as good as those above.

Charles likes pressure treated 6x6s

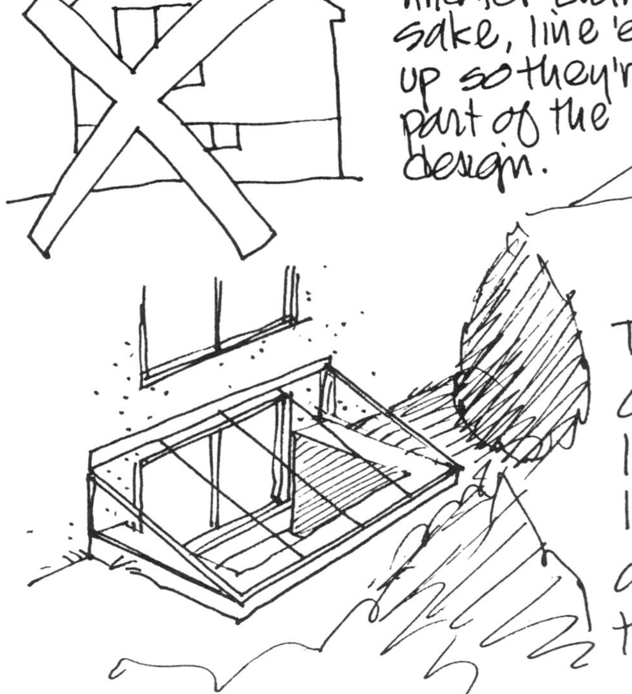

And for Evans' sake, line 'em up so they're part of the design.

There are many ways of dropping the grade level to accommodate larger basement windows and making them part of the design.

Good solar practice applies to basement windows, too. Unless you're in hot country locate the larger windows on the sunny side.

The no-triangles-in-doors-or-windows rule. No diamonds, either.

"Unless", adds Charles, "you are a post-modernist, which I hope you are not!"

I don't know why Charles thinks we have to mention this rule. You wouldn't do a thing like this, would you? And yet he claims to have seen people sneak these little goodie's into their houses all over the place down there in Pennsylvania. Up here on Cape Cod they do something far worse: they install nice ordinary double hung windows, then they come along and add snap-in divisions! Made of plastic, too! It's true. I've actually seen them.

← picture of a common storm door Charles hates. I eschew such things also.

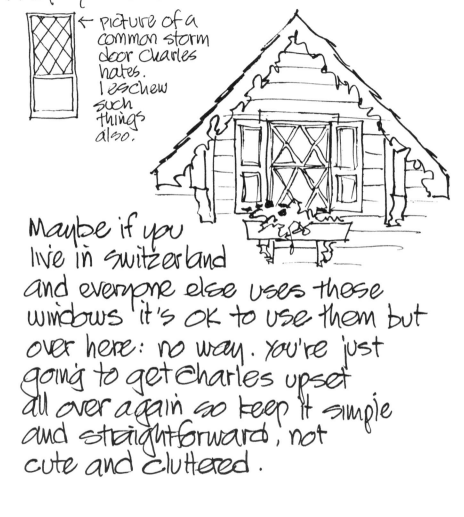

Maybe if you live in Switzerland and everyone else uses these windows it's OK to use them but over here: no way. You're just going to get Charles upset all over again so keep it simple and straightforward, not cute and cluttered.

Stick to the basics when you think of windows ... admitting winter sunlight, preventing winter heat loss, durability, maintenance, ventilation ...

Watch out for overly ornate exterior doors, including garage doors!

If you or a friend is a talented craftsman, go ahead and put some special touches on your door, but then don't turn around and hide it with a cheap screen door from the lumber yard. Aluminum screen doors are often the gaudiest but the wood ones are no prize. You can upgrade them, both visually and physically, like this:

thin boards or plywood

To which Charles adds, "don't leave plastic doors gray paint/stain the extra maintenence is worth it!" He talks fast, too, so you have to pick up the ideas quickly. You get the idea: we're overloading this poor continent with trash – tasteless throw-away garbage – and it's got to stop.

Modifying standard doors is inexpensive, satisfying, and successful. Turn the page for some modification tips...

Charles adds: You can get nice insulated brown aluminum doors.

The Brits call insect screens bug nets. By either name they do gently what pesticides do violently. No chemicals. No toxic pollution.

Mac's ideas

There are innumerable ways to make doors more like parts of their houses, or to enrich their designs without being gaudy.

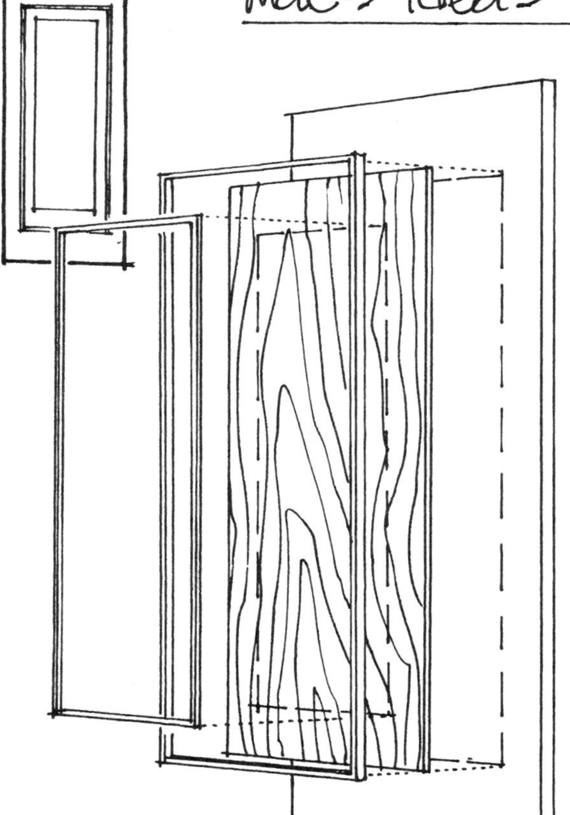

This is a low-cost hollow-core door to which 2 strips and a thin sheet of plywood are added on each side. The result: a handsome, solid-sounding door at low cost.

Here, a Dutch door is cut at exactly the height of the masonry. The cut, of course, is a stepped one with full weatherstripping.

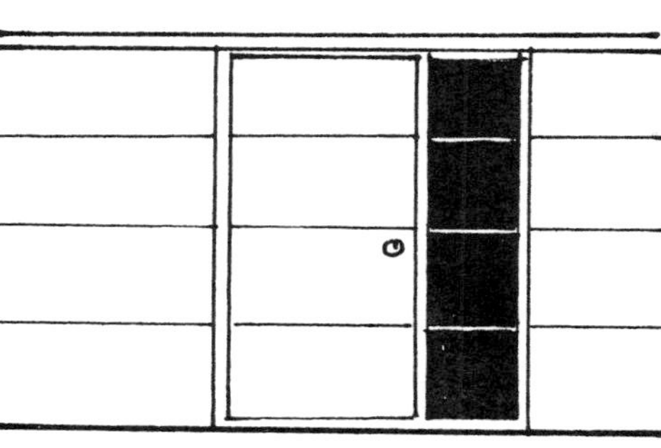

Carrying the garage door divisions right around the house, using little strips or sawcuts, or even, as on the glass sidelight, 1/8" stripes of glossy white oil-base enamel.

Why am I not outside planting another tree instead of making you read all this gray matter?

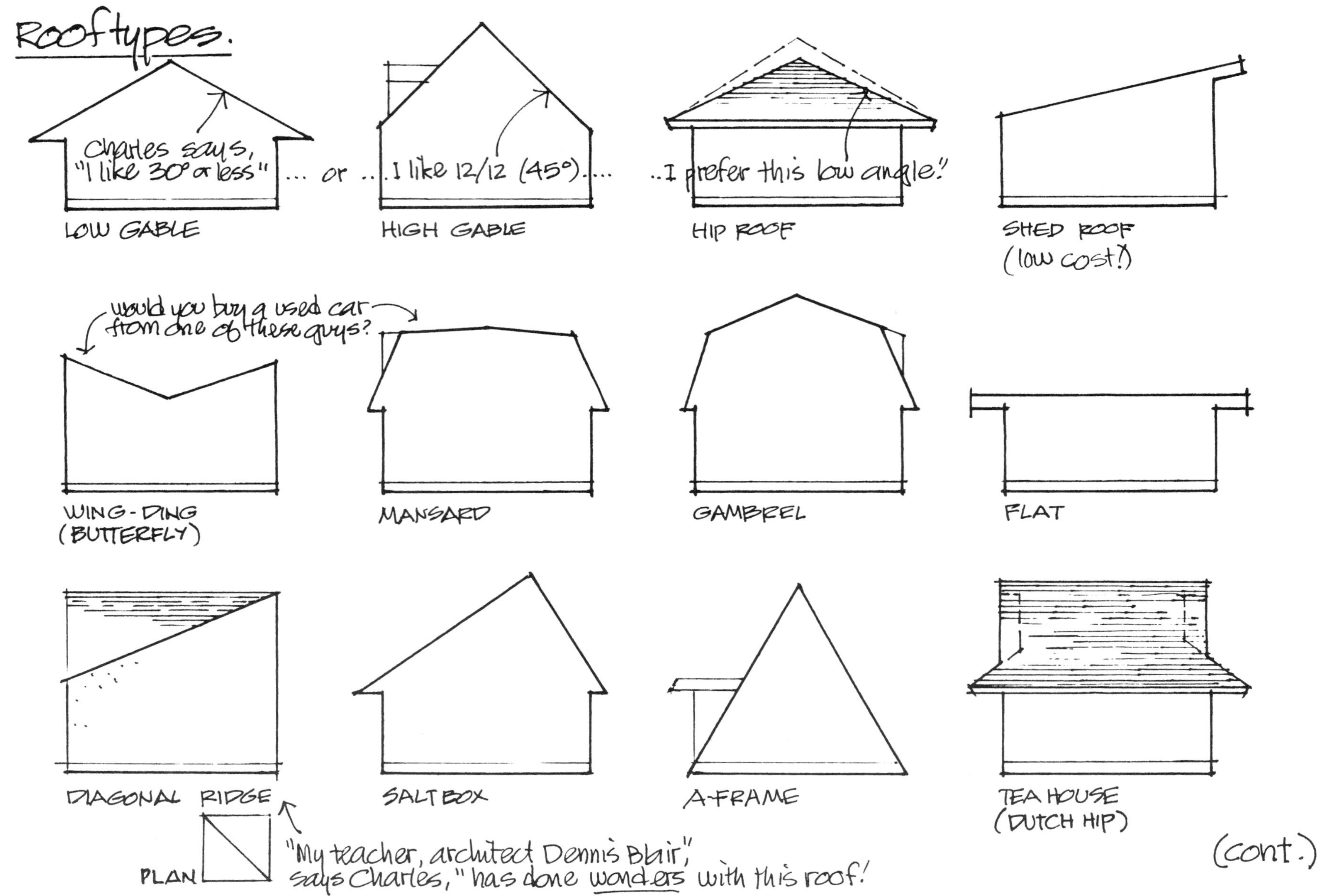

Now you need no longer wonder why we Americans are so wasteful and spoiled. Look at all those choices ... as our forests disappear.

So just what *is* the rule for selecting a roof type? Whim? Keeping up with the Jones? Or is it the emerging sense that we must be responsible for whatever we do on living land? ... that our choice will affect the lives of plants and animals far from where we build?

If we don't build permanent buildings the natural part of our world will never have a chance to catch its breath.

There are wooden houses in America that approach 400 years of age but they are extremely rare. And the average house built today is a lucky one if it stands for 50 years.

Architecture isn't a paper handkerchief. The throw-away era must end. Please try to build long-lived structures for the sake of the earth and of the seventh generation.

Use simple roofs unless you are an experienced designer. Simplicity usually looks best, and it conserves natural resources.

Create lively spaces under the roof.

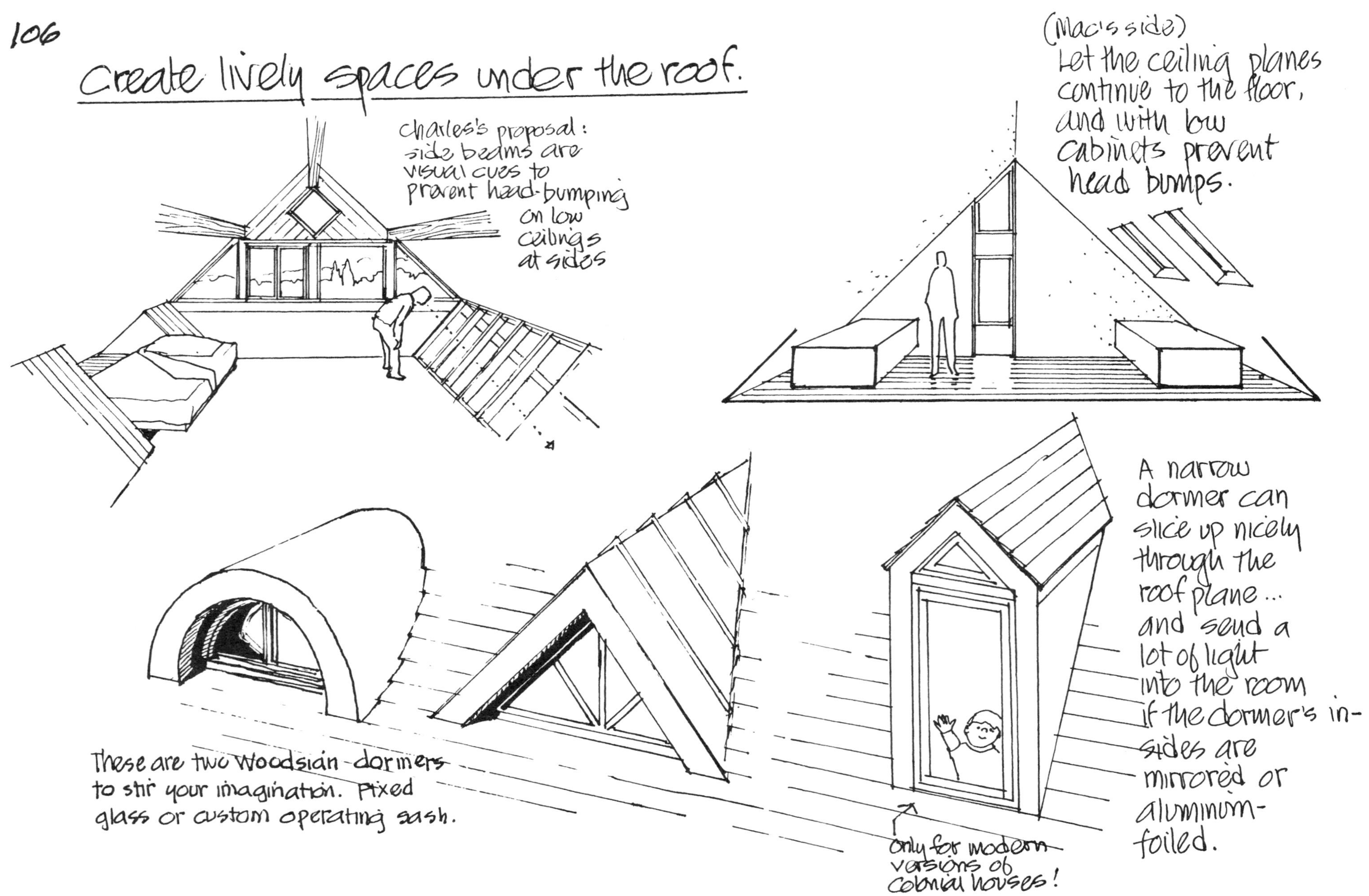

Charles's proposal: side beams are visual cues to prevent head-bumping on low ceilings at sides
(Mac's side)
Let the ceiling planes continue to the floor, and with low cabinets prevent head bumps.
A narrow dormer can slice up nicely through the roof plane... and send a lot of light into the room if the dormer's insides are mirrored or aluminum-foiled.
These are two Woodsian dormers to stir your imagination. Fixed glass or custom operating sash.
only for modern versions of colonial houses!

A lot of heat can come in — and go out — through the roof. Moisture can be a problem, too. Build it right.

Use the same roof materials - and angles - on all parts of the house.

Oops! Well, I guess that's the end of our famous cover house. It seems that it breaks this rule of roofs.

So, just as an interim measure let's see how much better the house would look if its roofs were brought into line with the Charles Woods Rule of Roofs. Quite a change, isn't it? And we've hardly even gotten started yet. Stick around.

The environmental consequences of roof-design unity? Take a look. See how much better the landscaping has become? Voila!

Use wide roof overhangs

Can you imagine standing in the rain at the front door of a no-overhang house, waiting to be let in, or waiting till someone finds a key? It's a dumb design, and it has many other weaknesses as well.

Without a proper overhang a house will get wet inside when windows are left open in the rain, and all the building materials left out there in the weather with no protection from sunlight and rain will deteriorate much faster without a sheltering overhang. So don't be timid; express the idea of shelter with 3, 4, or 5 feet of overhang... whichever best suits the lines of the house.

Mitered (no mullion) corner glass is nice in mild climates

Now you have a house that looks inviting, protected and secure. That's what houses are supposed to be.

Unless it's always cloudy where you live, size your overhangs to admit the low-angled sunlight of winter and block the high-angle rays of summer.

Expose as little flashing as possible.

Flashing is that almost-hidden metalwork that keeps roofs and windows and the tops of doors from leaking.

Charles wants as little of it as possible exposed and that's OK as long as the gap between shingles and wall, for instance, isn't so narrow that air can't get in freely, allowing the materials to rot. You can conceal much of the flashing with paint so that no twinkly shiny aluminum exposes itself, or you can use pre-finished aluminum in a color compatible with your other materials. Lead and copper sheets, the traditional flashing materials, weather to appealing appearance but they're being priced out of sight, giving way to aluminum, and, more and more, to plastic.

Protect the house from leaks and rotting, they're the main things.

If the life of a building is doubled, its share of our dwindling natural resources is, in effect, halved. Quite a worthwhile accomplishment.

Integrate roof vents.

I'm sure you know of the plumber (who shall be nameless) who never even bothers to cut off the vents once they go through the roof... and who always locates them in the most prominent place. Well, that's it: we've had it! Old Nameless has been fired.

From now on all rooftop vents, covers, and other devices will be in the least conspicuous places. The vents will be cut off at the regulation 18" lengths, and they'll be painted roof color as well. You'll need a radar gun in order to find them. I mean why build a nice house and then cover it with metal and plastic distractions?

At my underground office the big tall vent of the composting toilet is painted the same greenish black-brown of the pine trunks beyond it and it is almost invisible.

—M.W.

The natural world needs—and has—as many vents as we and our buildings do but look how beautifully it conceals them.

I like heavy fascia.

That's a direct quote. I've checked and double-checked my notes. He definitely didn't say fasciae, so he must be talking about one particular fascia down there in Pennsyl-tucky.

treatment I like to employ now and then when no gutters are involved

"Instead of this.. says he. this.. or even this."

So he definitely is a heavy fascia man: "It helps emphasize the horizontal." But, for all that, he's open-minded. I asked him about a zero-fascia

and he was very kind in his response to it, so maybe the rule is to do whatever you do boldly with no ambiguity. Strong, clean design. Right, Charles?

"I use both but I prefer fasciae angled to the ground." →

Build it to last. That's the most important consideration. The wood of which it is built may have taken 200 years or more to grow.

Do not use fake roofs!

Charles, "what's a fake roof?" I asked. His response was this series of four little sketches from which I've deduced that he's talking about ceilings. Let them follow the roof line, says he.

"A roof full of trusses is totally for show and the storage room is minimum".

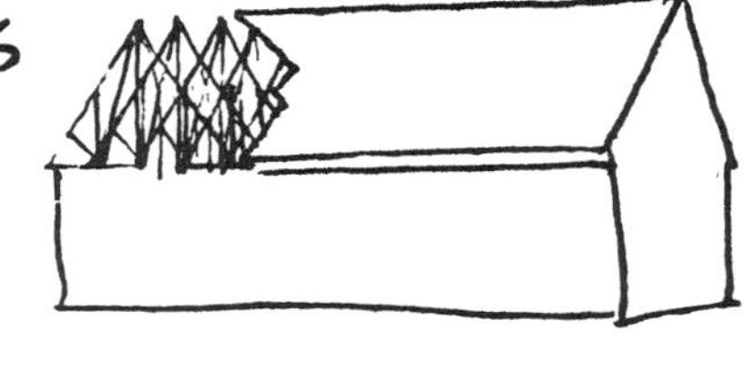

He goes on to say that if you want a flat ceiling then use a flat roof.

"I do not believe", he continues, "that trusses save much money," referring to the widespread practice of spanning wide → spaces with a network of small members.

Even when using trusses, use or (high ceiling ones)!

"But they leak!, people exclaim. "Not with a good membrane roof – butyl, etc. Then they are the least expensive of roof shapes," Charles replies.

So that's the story on fake "roofs."

And be sure to install water-saving faucets and shower heads.

Use double shingles* every 4-8 courses... or add a color strip.

Here's another of Charles's dandy ideas.

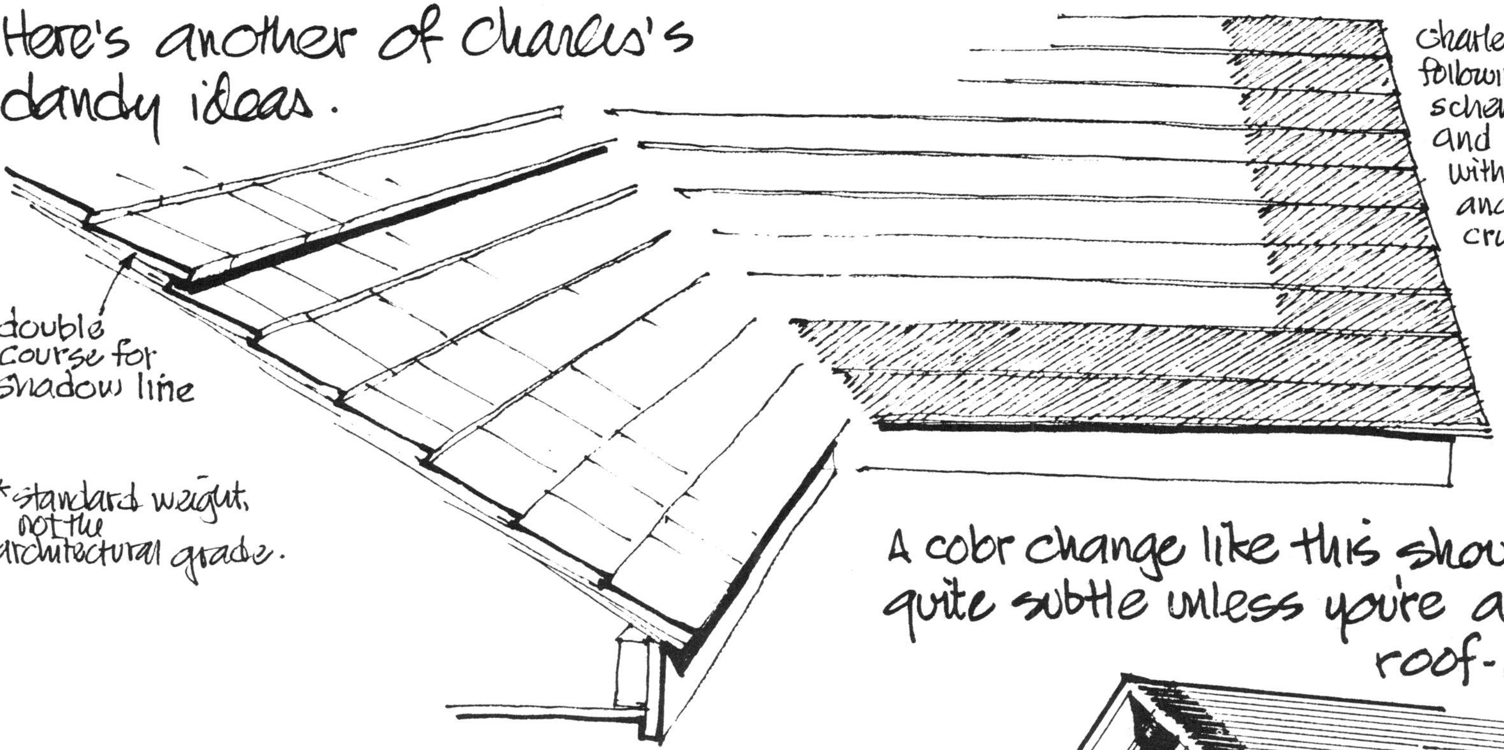

Charles offers the following color schemes: dark blue and red, or red with brown/green, and adds the cryptic, "I think."

double course for shadow line

*standard weight, not the architectural grade.

A color change like this should be quite subtle unless you're a master roof-colorer.

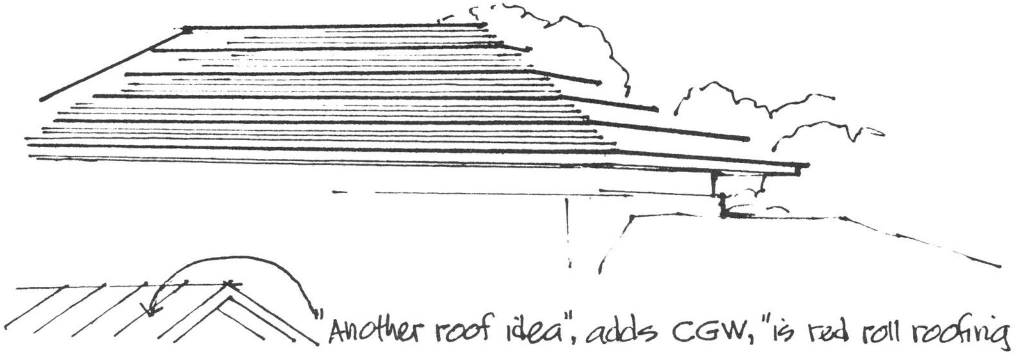

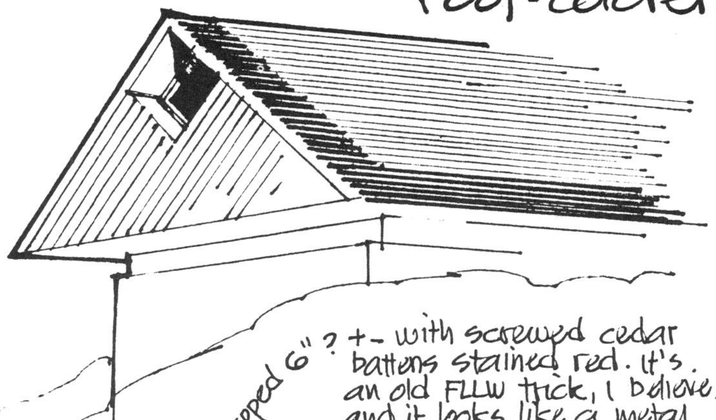

"Another roof idea", adds CGW, "is red roll roofing laid vertically and lapped 6"? +- with screwed cedar battens stained red. It's an old FLLW trick, I believe, and it looks like a metal roof, sort of."

Only one type of roof changes color with the seasons and grows healthier each year. Who can tell us what kind of roof that is?

"Color your roof!" say Charles.

"There are too many black or white roofs out there," he says. "Even brown is overdone. Try dark green or reddish."

(See the kinds of problems you create when you don't build underground?)

"Knock it off, Mac."

"That's Mac from behind," says Charles, who's never seen me. I'd have to add 100 lbs. of muscles to my skinny frame to look like this. I thought I was drawing Charles!

White roofs are much cooler in the summer, black better in winter. Try to select the tone best suited to your climate. Do you spend more on cooling or heating?

Ridge vents

One of the best on the market, by Air Vent, Inc. of Peoria Hts., Ill., looks something like this. Its louvers, baffles, and filter material are said to prevent snow and rain from blowing into the attic space. That's a major consideration. But I (M.W.) would like to see ridge vent caps with a bold profile.

It seems to me that a design like this might do all that the Air Vent design does, and offer in addition a strong ridge line that could even be run past the end of the gable a bit.

"Cool, Mac! Looks sort of Japanese! Copper would be real (sic) beautiful!"

If you help your house breathe properly it will reward you by living longer.

Use natural finishes and earth-tone colors with modern designs.

Don't use bright primary colors except in great overspilling flower beds massed around the entrance area.

Look to the natural world for a key to your pallette. but don't get carried away by a day in May: Charles says there are about six houses on one street in his town and they're all painted lime green! He says that God doesn't use that color very much so why should we?

Even for Colonials, use the traditional white or the muted colors on display in the Williamsburg color samples that are sure to be on display at your hardware store.

He – Charles, not God – goes on to say that you shouldn't paint garage doors bright yellow, or house trim red.

ugh. who would do that?

Latex paints seem to cause far less environmental damage than solvent-based paints, and in my (M.W.'s) experience now last just as long.

Use as few different materials as possible.

Don't mix different masonries or sidings outside the house, and try to restrain yourself inside, too. Architects are as guilty in this regard as anyone. Too many of us try to show off our versatility by specifying as many materials, textures, and colors as possible.

But look at the classic buildings of antiquity. How many materials in each one? One. Maybe two. So why do we think that more equals better? We're spoiled, that's the trouble; it's too easy.

It's not at all unusual to see a house with brick, stone, and block masonry as well as aluminium siding, wood siding, concrete steps, iron rails, and aluminium windows. With plastic shutters! How the gods of design must cringe at such shenanigans! And yet such shenanigans are generally accepted as the norm during this ongoing binge of materialism.

Well, we're going to put an end to that right now, right, Charles?

Right, Mac. You bet we are.

Every material used has a direct consequence in the natural world. Water pollution, energy waste, and habitat destruction are tied to every product.

Use stucco.

It got a bad name back in the Thirties when they used to trowel the cement onto wood laths and it would crack and fall away in big chunks after a few frosts, or after the strips had rotted.

Now we use sheets of galvanized metal lath, nailed (with galvanized nails) to the structure (if the frame is wood). If the house is built of block masonry the rich cement mixture is spread right on the block. Two additional coats conceal all the block joints and make the surface last.

Standard cement finish colors are gray and white. If you want other tones, such as yellowish tints for warmth, Charles likes to dye the mix. I've never had good luck with dye, finding each batch a different shade. I use full-strength or sometimes watered-down latex house paint. If the stucco has been left rough (as I prefer it) the color holds for decades. Just be sure to avoid the repeating trowel patterns plasterers love when creating a rough finish. Make it random, patternless.

A good stucco job seems to last indefinitely and require no maintenance. That means fewer refinishing materials gassing their poisons into the air.

Avoiding cracked tile on wood floor construction.

Until Charles said it was OK I thought that the use of brittle, earthy, masonry-like materials on springy, shrink-prone material such as wood construction had been banned at the Geneva conference but he says that as long as you do it right, preventing cracks by beefing up the substructure, there's nothing at all wrong with the practice, so you listen to Charles, hear?

dyed grout joints

earth tones

tile

"Durorock" (or equal) setting bed

thin slab of mortar on galvanized metal lath nailed to tongue + groove plywood subfloor. stagger joints.

floor joists 12", not 16" or 24" on center. Usually 2x12 Douglas fir.

12"

Douglas fir. Limit spans to 14' ±.

"Many people build over a basement, Mac, so we have to consider them."

Dark-colored tile is best for absorbing solar radiation and releasing it to the room after sundown.

Thoughts on flagging.

Like almost every other material, flagstone can be badly misused so it gets badly misused often.

Unless your building is blue or red resist using red* or blue slate flagging. Gray and gray-green, and where available, earthy browns, are best*. Let the floor or patio be richly understated like the earth. Flowers, drapes, books, and art can supply all the color you need.

Charles: "I like red but not blue usually." Aha.

Charles: "I agree, though."

"But dye the grout," adds Charles. "Light gray shows up all the imperfections."

So, first of all, don't use any concave shapes. That was easy.

Now:

Whether you use irregular or cut rectangular flagstones, keep a straight architectural edge on at least one side.

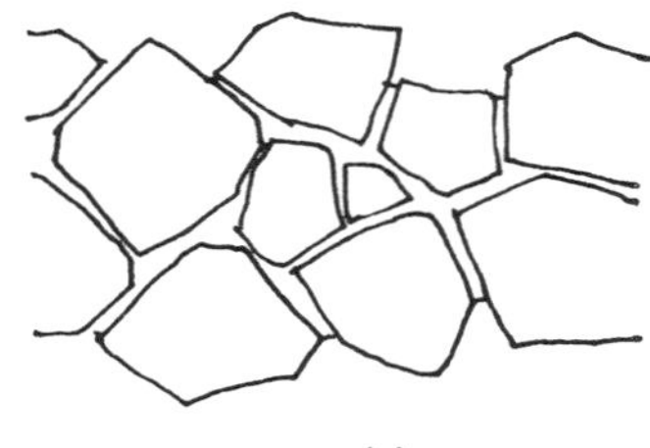
No

yes

yes

yes

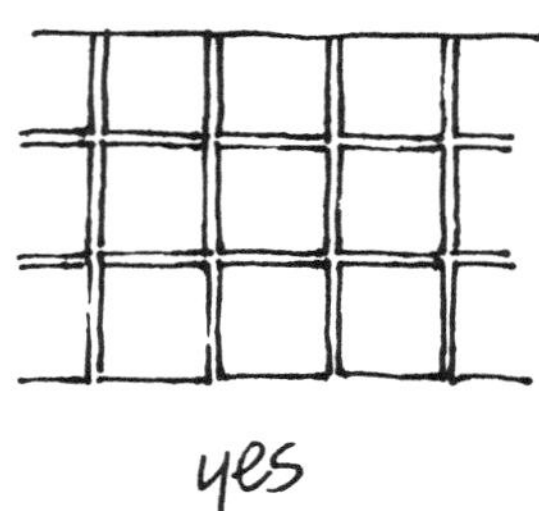
yes

Stone, like concrete and tile, feels cold underfoot because it conducts body heat away so fast. See "sniffer ducts" for a way to take the chill away.

Stain all pressure-treated decks the same as the house.

"What if the house is all brick?", you ask. Hmm... Charles, what about that?

"Then build a brick patio!"

Okay.
The idea here is unity, continuity, reducing the clutter. Let the house and deck, visually, be a single entity. An alternative to this rule, of course, is to stain the wooden parts of the house's exterior the same yellow-green as the treated wood. It all depends upon the color scheme you choose to go best with the colors of your landscape, particularly in the winter when any non-natural color scheme will leap out from among the muted grays and browns of the season and smite you between the eyes.

try to conceal all supports.

You might also consider making not only the colors but the <u>lines</u> of the deck be a part of the house.

And remember, says Charles, you can add house siding to parts of the deck rails.

Be careful of that poison-treated wood. Read the fine print about breathing sawdust particles, washing your hands, etc. Be skeptical of all toxins.

Textured or sand-finish paint helps improve the look of drywall.

charles probably feels this way because they don't know how to finish plasterboard properly down there in Pennsylbama. It's the only way to cover all the joints and nail-heads, I'll bet.

Up here on Cape Cod, they do such a good job you can hardly

see the studs afterward.

In any case you should listen to Charles because he has an eye for these things. Just don't make the paint too heavily textured. When you sight across the surface of the painted wall you should just be able to detect the ups and downs of it. If they project too far all the subtlety will be lost.

"Use sand/sponge finish," says Charles.

Every ounce of weight you add to a house increases its thermal stability but that doesn't necessarily mean warmth. Design carefully!

Charles Woods on Interior Decorating.

I haven't said much on this. It may become the subject of another book. But I did give some hints about it – simplicity, consistency, honesty, etc.

It unfortunately only takes a little French Provincial furniture in a modern house, or a black-light Elvis in a carved gold frame (sorry, Mom!) to wreck any house – at least for magazine publication.

How do people have individuality then? Don't little girls get pink rooms and little boys blue? Doesn't your teenager get to paint her/his room bright red or black, etc.? (I did!)

Can't a wife have those blue bathroom fixtures? .. or the husband use the octagonal tile he found in the store, etc. etc. etc.?*

This really is a tough problem. My suggestion: try it the architect's way for a while. Paint the whole interior ivory and use one tile color throughout. Pretend you're doing it in order to impress an editor of a home magazine. Then see if the colors and personal effects you want can be brought out in details. For instance, a large corkboard wall might satisfy both your architect and your teenager's poster-hanging desires! Or you might satisfy your desire for bright red in the bath by a beautiful

*Let the husband use the octagonal tile, I say. I mean what's the big deal, one lousy tile?

Interior Decorating (cont.)

Japanese or Mexican vase, etc. After a year or so if you're not happy paint the rooms the colors you want. Paint doesn't cost much. Go paint all the rooms pink and blue. Our individuality is important. But what individuality, self, or Self?

Herman Hesse, one of my favorite novelists, pointed out, I think in The Glass Bead Game, that true individuals tend to look more, not less, alike. Jesus, Buddha, Plato, Joan of Arc, LaoTzu, Gandhi, King, etc. do seem a lot more alike with their real individuality than our teens and movie idols with their ever-new haircuts (and I've tried them all!) and the latest clothes.

What does this have to do with interior decorating? Something but I'm not sure what.

Fireplaces should be in the center of the house*

Fireplaces - and chimneys - are very wasteful of energy, especially if the chimney is on the outside of the house. Even when not in use chimneys are heat-leakers. Room air slips past the damper and out it goes. "Russian Fireplaces", which are huge masonry affairs that wring all the heat out of the fuel by making it zigzag through the mass of masonry, are efficient. So are the more recent woodstoves. But all burning of wood and coal adds to the overdose of CO_2 we're pouring into our skies.

The fuel of first choice must be sunlight. Sunlight, and insulation to conserve it. Ah, but isn't a fire appealing? The glow of embers, the crackling flames, the warm masonry in the heart of the house... their magnetism seems to touch something primitive in us.

Is a compromise possible? Of course: an efficient-as-possible fireplace used only on special occasions - not to heat the rooms but to warm the heart. Make sure the damper can be closed tightly afterward, swear you won't use your car tomorrow (by way of compensation), and let 'er rip.

Now for the Charles and Mac designs.

* "The symbolic center of the 'soul', of 'life', and of 'light', traditionally," adds you-know-who.

Can fireplace ashes be safely used to enrich the garden? The answer used to be yes, but now, with all the toxic materials in the kindling paper it's a tough question.

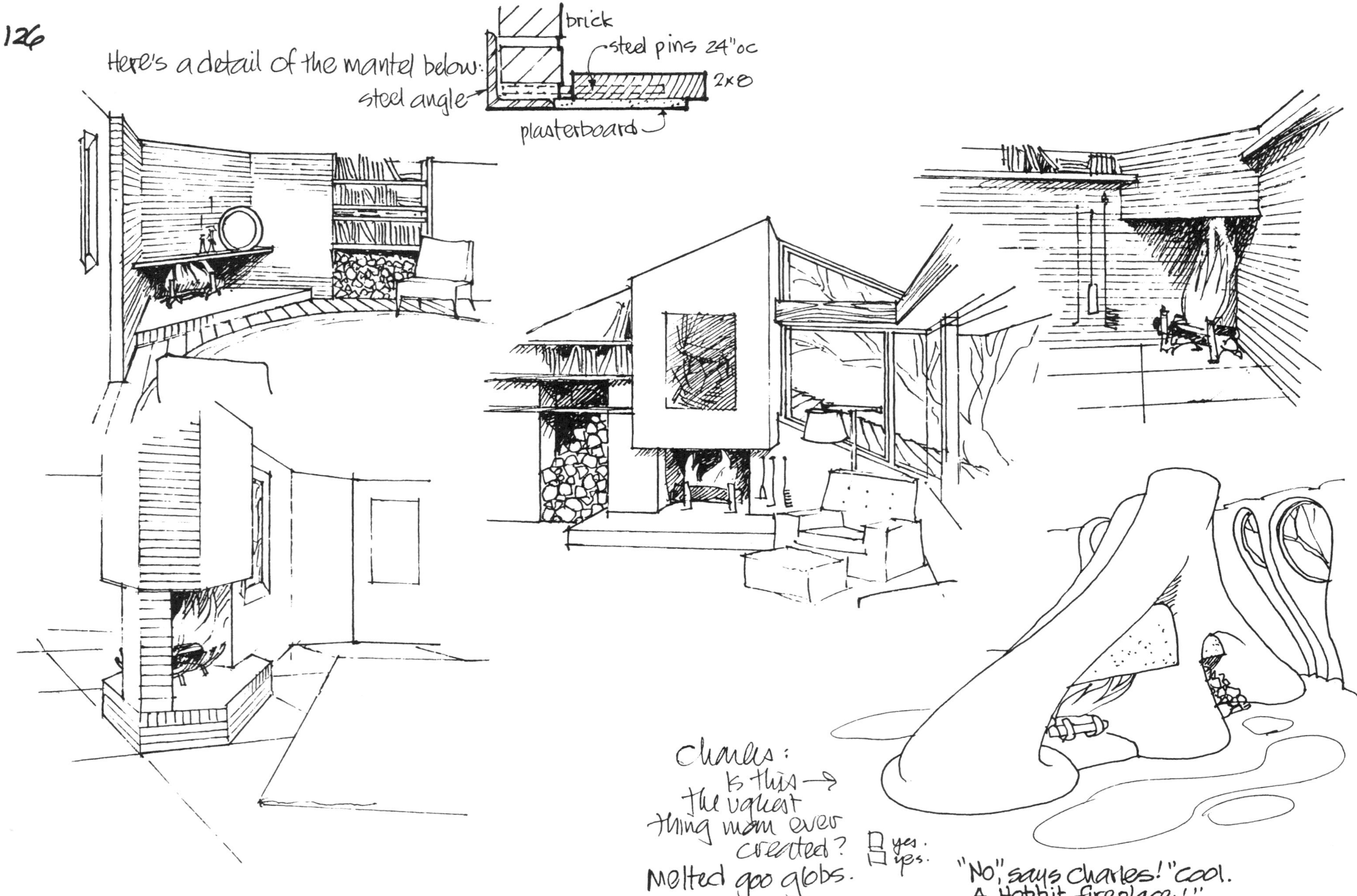
Here's a detail of the mantel below:
brick
steel pins 24"oc
2×8
steel angle
plasterboard
Charles:
is this
the ugliest
thing man ever
created?
yes
yes.
Melted goo globs.
"No," says Charles! "cool.
A Hobbit fireplace!"

Chimneys can also be fitted with operable flue caps (operable from indoors) to further prevent unwanted heat loss.

Make chimneys massive – and keep them toward the middle of the house.

Brick, block, and stone look – and act – unstable when used in tall thin structures. Chimneys should look – and act – stable and permanent; the cores of the buildings.

Exterior chimneys lose valuable wintertime heat to the out of doors. A chimney in the heart of the house radiates its warmth to the spaces that need it.

Somewhere along about now you'll be itching to say, "Whoa! This is all good stuff but it's going to cost a fortune!," and you'll be exactly right. If you did everything we recommend it would add a lot of cost to your house. It would also add a lot of value, too, but that's not much consolation. The point of the book is to get you to use the principles, not all the details, and to stay withing your budget.

charles would like the record to show that big chimney near the center of the gable end still looks OK.

The goal, of course, is to have no chimneys but if it's achieved by using electric heat the pollutants are simply moved to the power company.

Put tops on masonry chimneys.

Topless

The usual chimney is left unadorned, and Charles would like to see some tops like these. Isn't he a whiz?

dyed cement to match grout

provide clean out cap with handle

stucco'd over wood

siding

I'm not so sure how certain caps affect the draft of a chimney but I've always had good luck with my big flat-top caps so I guess they work all right. At least they give you a chance to repeat your architectural statements up where all the world can see them.

And of course there's nothing wrong with the old standard chimney top. Just don't leave the poor thing naked like this:

Or worse yet: concrete block, usually dirty

Geometric holes must be large enough for proper draft!

Does every chimney – and exhaust pipe – say "global warming" to you?

Add boxes to metal flues.

Charles feels very strongly about this:

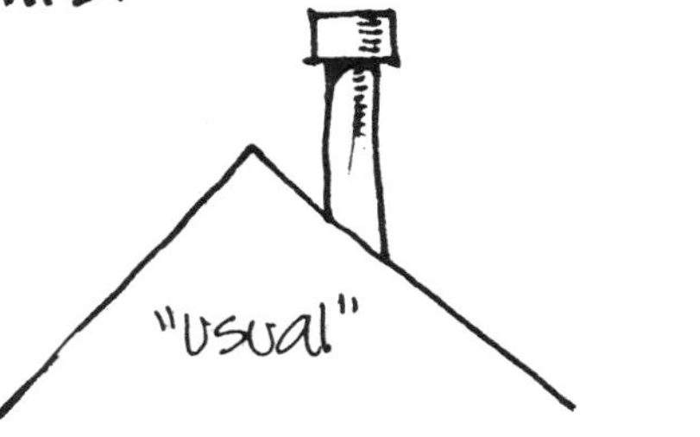

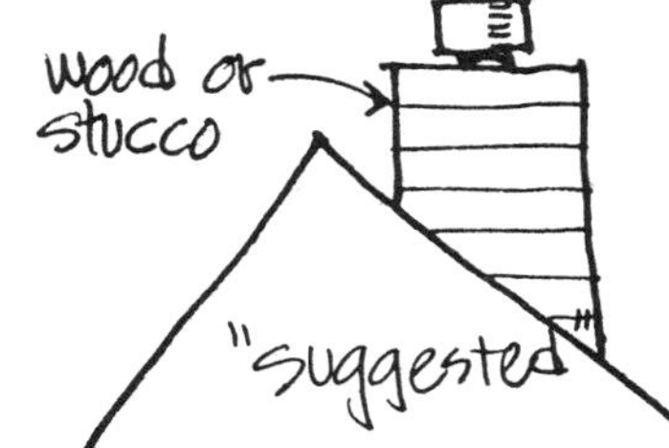

So do I:

Down with fake chimney enclosures

"yes, Mac, OK," says Charles, "but what about this? That's the problem.

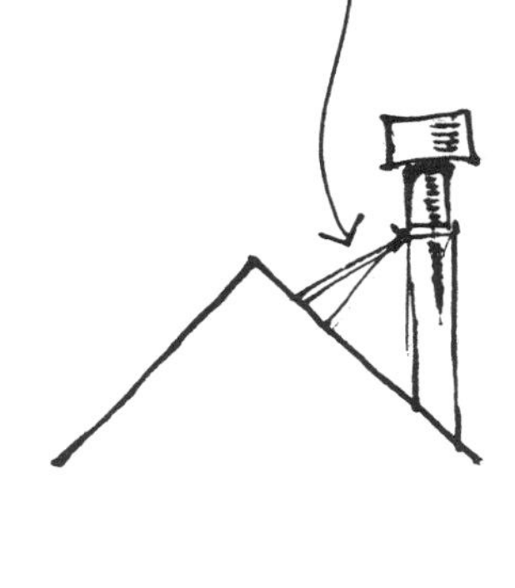

It just seems to me that even though this isn't very pretty it isn't right to enclose a fire device inside a wooden box, no matter how unwoodlike it's made to look. Let it be what it is until you can afford a chimney made of materials that take fire in their stride.

"If you must expose this make it a good brand like Metalbestos". —C.G.W.

We have fun disagreeing on things, Charles and I, don't we Charles?

"yes."

"Is boxing-in a metal chimney dishonest?," asks Charles, and he answers himself with Emerson's words: "Consistency is the hobgoblin of small minds," and then Charles adds, "but be inconsistent only when Charles tells you to."

MANAGING ENERGY

Big considerations: energy efficiency, south-facing glass with properly-sized overhangs, high-efficiency glazing, heat-holding windows, earth berms, super-insulation, photovoltaics*, composting toilets, tiny lawns, porous paving, organic gardens, convenient recycling bins, and, above all, responsible occupants.

*electric power, direct from solar panels

This page is so good, environmentally, that the authors have awarded themselves medals.

Complete the insulation envelope without gaps.

Insulation not only does the obvious jobs of keeping winter heat in and summer heat out, it has a lot to do with moisture control, too, especially in the summer. Floor slabs on grade, in particular, and basement walls should be insulated on their earth-side (outer) surfaces to prevent the slabs and walls from being cooled too much by the surrounding earth. Warm, moist air will condense on cool surfaces, and that's not good.

Under some conditions, of course, no amount of insulation will provide relief from hot weather. That's when air conditioning may be justified. If you use it, be sure to use the smallest unit that will do the job when working at full capacity. Not only is it most efficient when worked hard, it's also the best dehumidifier, too, getting a lot of moisture out of the air before its thermostat is satisfied.

If you live in the southwestern desert you'll do well to find a local expert in these matters.

And the underground man says your house will avoid the extreme highs and lows of outdoor temperature if it's nestled snugly into the earth.

It is important to understand what vapor does in walls and ceilings.

Heat always moves toward cold. When you feel a chill you don't feel coldness entering your body. You feel heat leaving it faster than your internal heater can replace it.

Similarly, the water vapor that's present in all air will move from a warm place toward a cold, or cooler, one. Typically, in the winter, the moist air will try to move toward the out of doors in any way it can. That means through the walls and through the roof. Walls may look smooth and airtight but to water vapor they look like sieves, and out it goes.

That can cause a big problem because whenever that moist air gets cold it drops its moisture as condensation. (Water.) So unless you stop the air from leaving the warm room it will drop its moisture at some point about half way through the wall insulation, turning all that nice woolly fiberglass into useless soggy mush, and ultimately rotting part of your wall.

So you install vapor barriers on the warm side of the wall, and seal every opening around doors, windows, electrical outlets, etc. Some of the air will still get through but the

A few dollars spent on vapor control can save hundreds, perhaps thousands, of dollars in fuel and maintenance costs over the years.

Want to see some of Charle's speedwriting? Look: ARINFL.* MRINFIL or IHRINHL?

sealed quality of the rest of the wall should allow that small quantity to evaporate harmlessly.

Heavy line = vapor barrier 6 mil plastic lapped 6"

moisture

insulation

insulation

The same arrangement holds true for the roof but in order to assure the needed evaporation of moisture in the insulation there must be a free flow of air from the eaves vents to the ridge vent. This is especially true in the summer when superheated attic air would otherwise drop its moisture, by condensation, on the relatively cool insulation and ceiling below.

Remember:

1. Tiny soffit vents in the eaves and little screened louvers at the gable ends won't do the job. Use continuous screened strip vents at the eaves and a snow-and-rain-proof ridge vent at the top, with color to match the shingles.
2. Maintain a clear air passage at least 3" high above the insulation and under the roof.

Amen.

---

*"AIR INFIL"(TRATION), he explained.

Air Vent, Inc. "Filter Vent" and "Shingle Vent II" seem to be the best ridge vents. 1-800-AIR VENT. Peoria Heights, IL 61614.

charles highly recommends it.

Are you ready to install a heat exchanger? ... a sniffer duct?

If you've done everything you should to make your house as air-tight as possible (as a way to reduce wintertime heat loss by infiltration) the level of pollution indoors could become dangerously high. You need fresh outdoor air but if you bring it in cold all your weatherstripping efforts were wasted. What you need is a heat-exchanger.

indoors

outdoors

wall

It's a machine that pushes thin layers of warm indoor air out as it brings thin layers of cold outdoor air in. The membranes between the layers allow much of the heat to cross over and ride back into the house on the fresh outdoor air. Heat exchangers are becoming standard appliances, easy to find from many sources.

—

A similar air/energy transfer device — one that merely recirulates indoor air, is often called a sniffer duct. All it does is pull wasted warm air that's risen to the top of the room down to floor level, either blowing it

can you spot this, my first mistake?

Until we use things like heat exchangers — and 75-mpg cars — we can hardly complain about oil spills and air pollution.

directly into the room at floor level or, better, into 4" tubes cast into the floor slab. Much of the heat in that room-top air will be transferred into the great mass of the floor slab, helping to stabilize interior temperatures in spite of outdoor weather fluctuations. A low-speed blower will use little energy and will be very quiet. And if a mouse should die in one of the tubes some cloth on a string can be pulled through to remove its remains for embalming, casket selection, funeral services, and proper burial!*

* How silly most human activities become when seen from another perspective! So much to do; so little time!

A domestic solar hot water system.

FIND THE BEST COLLECTOR-
TILT ANGLE FOR YOUR AREA,
FACE THE ARRAY DUE SOUTH.
MAKE SURE ALL WALL AND
ROOF PENETRATIONS ARE
TIGHTLY SEALED.

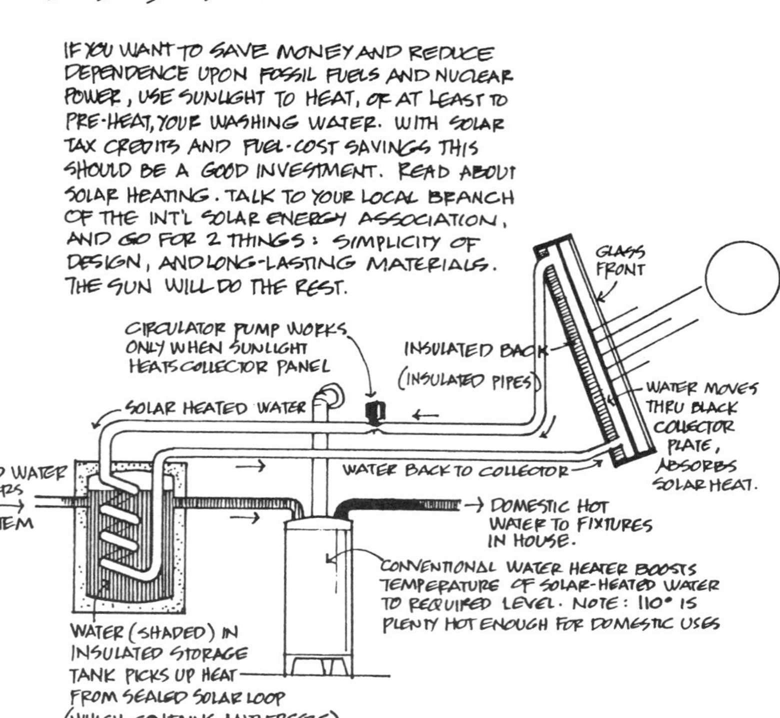

When the kids enter their bath-a-day stage, turn down the thermostat on the water heater. Turn them off a bit with 95° water.

Use indirect lighting.

Out of doors, if you keep all light sources hidden you'll be able to see stars instead of light bulbs. Don't use lamps and lanterns that blind you to the wonders of the night. Six or eight of these walk and driveway lights cost very little.... a few treated 2x8s, some sheet copper hoods, cheap porcelain sockets, burial-type wire, and 25-watt bulbs.

Twenty-five watts? Are you crazy?

Try them and see. Otherwise, your place will be lit-up like Broadway in a land that was always meant to be dark at night.

Indoors, indirect lighting will make the plane of the ceiling appear almost to float above you, and it will cast a soft, shadowless glow on everything else, making Charles glow, too, as he enjoys his designs, but it is a terribly inefficient way to light a room for specific activities like reading, sewing, crafts, or making sure you've getting those dinner dishes really clean.

The best indirect lighting is that which you get while seated, on a bright summer day, under a big old tree.

Payback on energy-saving features?

Larry Wilson, according to Charles, says that since energy-saving features often save 10 to 15 percent on fuel costs, and mortgages are in the 8-to-11% range it literally pays you to include the energy savers.

"But wait a minute", I say, "the 10-to-15-percent savings are on the cost of the fuel, not on the costs of the energy-saving windows, appliances, etc. on which you'll be paying 8-to-11-percent interest. Hmm... sounds like apples vs. oranges to me, Charles. Will you please go over this one again?"

"Still a 4% savings, Mac", replied Charles.

To which I say, "Gotcha (I think)". Look at these graphs of what you're saying. Is that more what we're talking about?

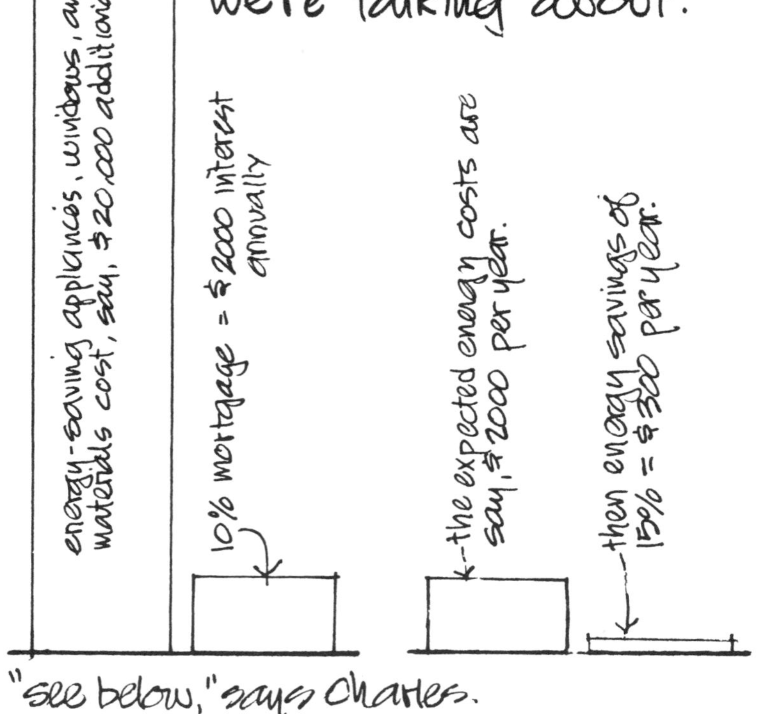
energy-saving appliances, windows, and materials cost, say, $20,000 additional
10% mortgage = $2000 interest annually
the expected energy costs are say, $2000 per year.
then energy savings of 15% = $300 per year.

"see below," says Charles.

If you care about the fate of the earth, of course, the dollar pay-back is not your only concern. You want to do what's right.

Here's another one from Larry Wilson*:

Consider using earth tubes.

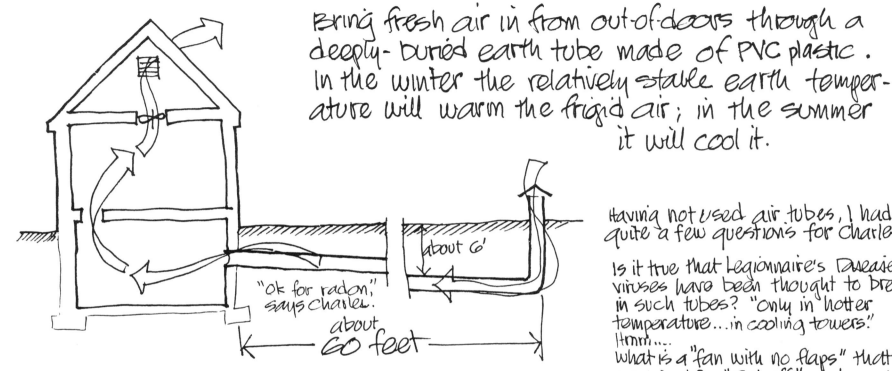

Bring fresh air in from out-of-doors through a deeply-buried earth tube made of PVC plastic. In the winter the relatively stable earth temperature will warm the frigid air; in the summer it will cool it.

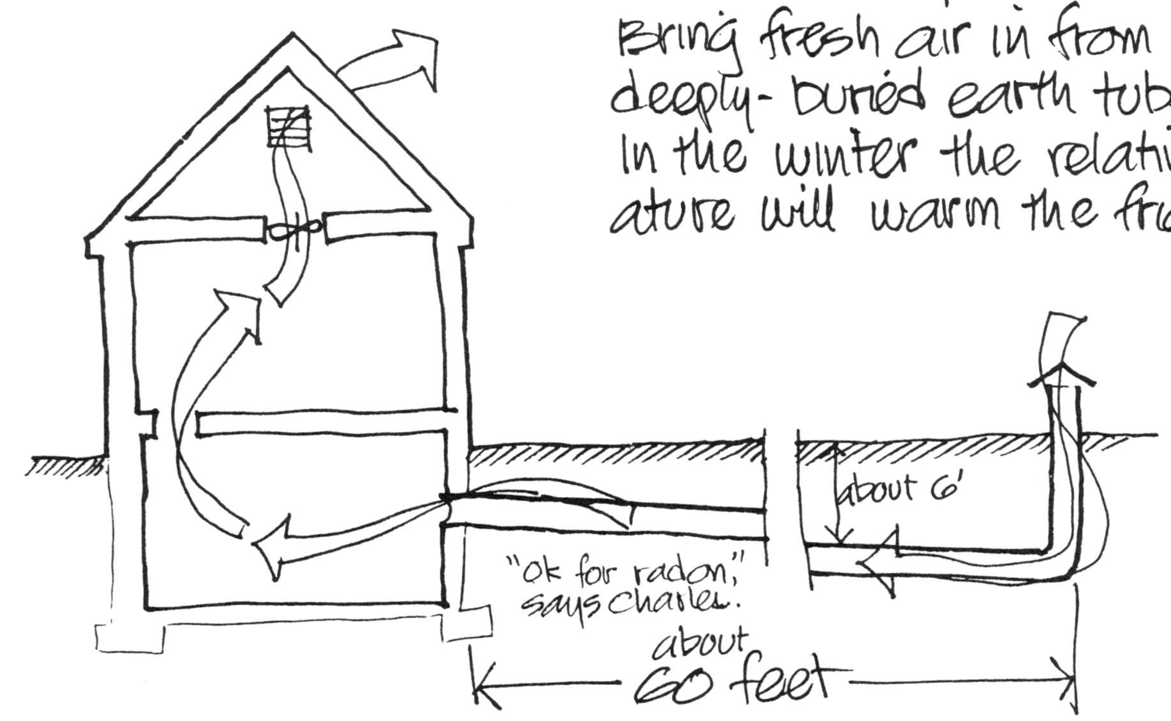

Having not used air tubes, I had quite a few questions for Charles:

Is it true that Legionnaire's Disease viruses have been thought to breed in such tubes? "Only in hotter temperature... in cooling towers." Hmm....

What is a "fan with no flaps" that you mentioned? "Cut off." Aha. huh?

How quickly does the soil around the tube reach outside air temperature? "I don't know, Mac. Larry's used the system, not I. It works!"

* "Larry Wilson is a general contractor, inventor, artist. He's the best builder I know. He's a genius. He taught me a lot about construction. He's also available for consulting through me."

I must say it sounds good. I've wanted to try it for years.

I hate to be skeptical but I think you should get good local engineering advice on this before trying it. And see if the engineer's ever seen a "no-flaps" fan.

"Solar attics really work," says Charles.

"I believe solar attics are probably one of the best energy systems we have."

Warm sunlight comes through the low-E insulating glass, warming the attic air, which is then pulled down into the rooms by a blower, as long as the thermostat tells the blower the attic is warm, after which an insulated damper covers the air pipe to prevent reverse heat flow.

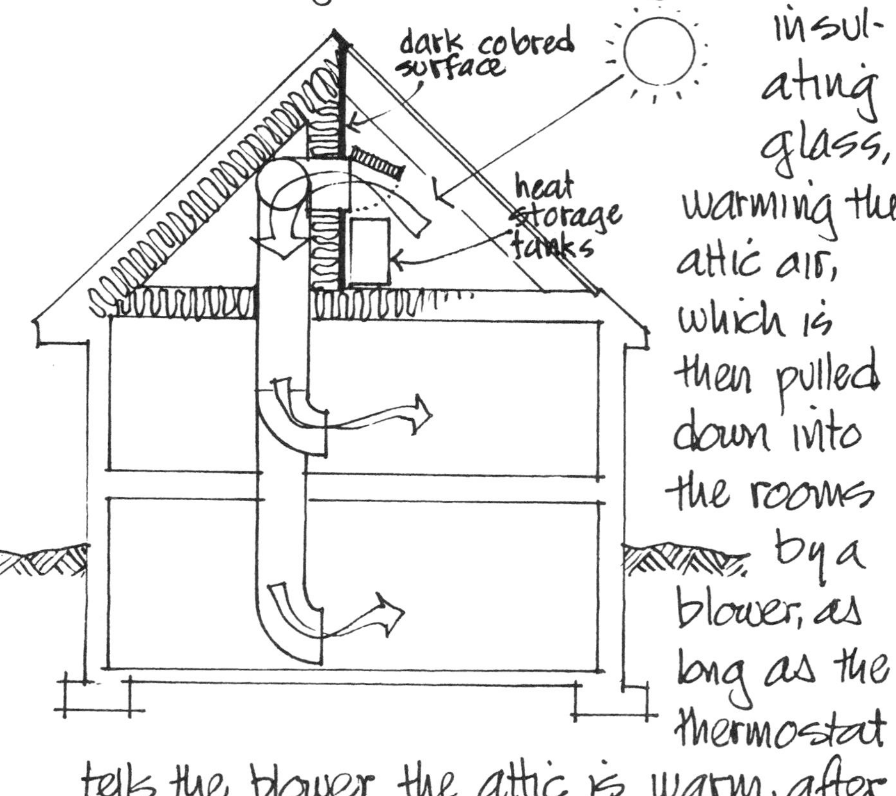

This also is a good space in which to house some hardy plants. And, says Charles, the <u>other</u> roof surface can be made to look quite traditional if that's what you want.

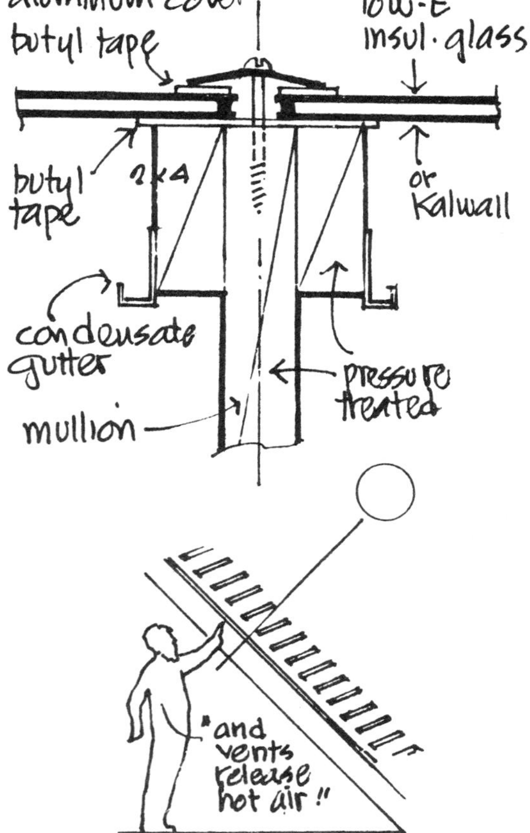

Here's Charles pointing to the exterior wood louvers he says can keep out the hot summer sunlight. Hmm... what kind of wood, Charles?

"Cedar."

Here's Charles's design for a solar attic:

And washing the windows?

"Rain."

And the skeptic says:
I'd give cedar slats about 3 years, max, in that murderous solar exposure.

Insulate basement walls on the outside.

Don't let water moving down through the soil touch the foundation in which so much heat-energy is stored.

Consider 24" rather than 16" spacing for joists, studs, and rafters.

The wider spacing and consequent greater depth allow for thicker insulation without reducing the 3" under-roof air passage so important for moisture control.

And he adds, "You must use high density 5/8" drywall for ceilings supported 24" oc."

Two rules on one page conserve paper (trees). Don't you feel less environmental guilt just being here to witness this?

Use modular layouts.

Always lay out the building, floor and walls, on a four-foot-square grid. This "modular" layout will save money and time by accommodating most standard building matirials. (Half modules, such as that of the closet above, work well, too.)

1 2 3 4 5 6 -
A B C D E
4' sq. modules

Most houses today are full of dimensions like 19'-3 5/8" and 7'-1 7/8" - and they look it. They seem to have no sense of order. They make you wonder just who was in charge of design, the designer or a shot in the dark.

Interior partitions fall on the centerlines of the modules; exterior walls call lie on the outsides of the lines or on the module.

Obviously, a lot of successful designs were done without modular planning but why make it hard for yourself? Why create unnecessary waste?

Charles says using modules will greatly aid design consistency and decisions - and window placement.

Waste means natural resources needlessly destroyed; buildings do enough land damage without having excess waste to deal with, too.

Also use vertical modules* (precut studs) for cost savings.

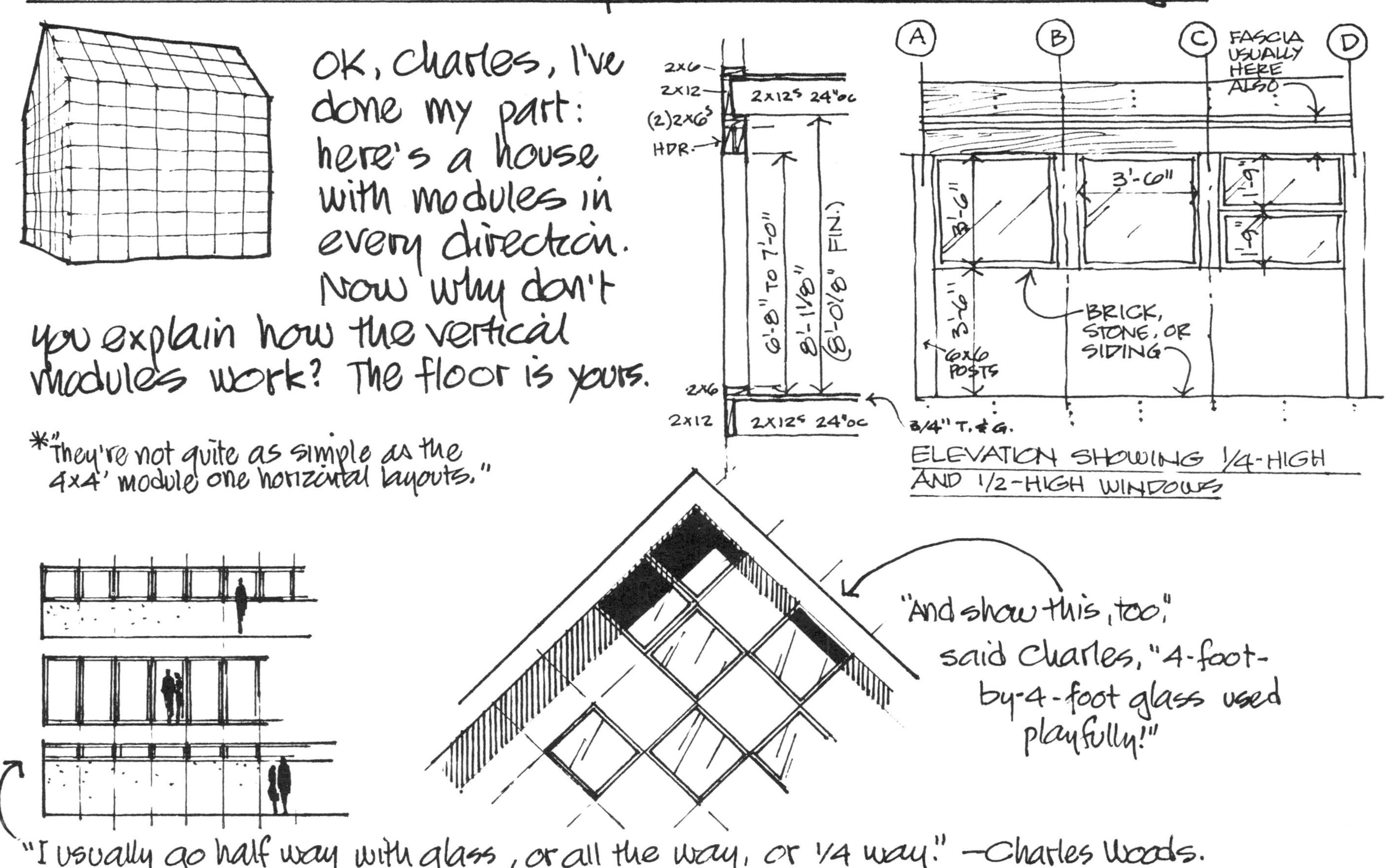

OK, Charles, I've done my part: here's a house with modules in every direction. Now why don't you explain how the vertical modules work? The floor is yours.
*"They're not quite as simple as the 4x4' module one horizontal layouts."
2x6
2x12
(2)2x6s
HDR.
2x12s 24"oc
6'-8" to 7'-0"
8'-1 1/8"
(8'-0 1/8" FIN.)
2x6
2x12
2x12s 24"oc
3/4" T.&G.
A
B
C
D
FASCIA USUALLY HERE ALSO
3'-6"
3'-6"
3'-6"
1'-9"
1'-9"
BRICK, STONE, OR SIDING
6x6 POSTS
ELEVATION SHOWING 1/4-HIGH AND 1/2-HIGH WINDOWS
"And show this, too," said Charles, "4-foot-by-4-foot glass used playfully!"
"I usually go half way with glass, or all the way, or 1/4 way." —Charles Woods.

I usually go 1/4 of the way in trying to make a house less destructive, environmentally. Someday I hope to go even further. — M.W.

More about modular, or grid, design.

Charles, you have the floor:

I have not always worked on modules consciously but looking back even to high school I believe I worked on them because I used so much continuous glass with posts every few feet.

I recommend using a 4-foot-by-4-foot square. This is the most common module used in construction. It nicely accommodates plywood sheets (4' x 8') and plasterboard (4 x 8', 4 x 10', 4 x 12'), and its best endorsement is that Frank Lloyd Wright used it much of the time. Five-foot modules are sometimes used, too, particularly in office buildings, where the 5-foot desk and the 2½-foot file set the pattern. Other good architects simply use structural bays of, say, 10 feet as their modules.

If you are <u>really</u> good you don't need modules. I'm pretty good but I still use them, and vary only when I have to. Frank Lloyd Wright went off the grid when he felt he should. Even God works playfully off of mathematical shapes and modules of various sorts.

4'x4 modules?

Pythagoras, the famous Greek philosopher, said, "God geometrizes," but forget metaphysics for the moment (even though my degrees, strangely are in philosophy).

Do you suppose Pythagorus segregated his wastes and turned off the lights when they weren't in use.

why do I use a module, and why 4-foot square? I find that it's hard to do a really bad house when designing it on the grid. Things just come out looking right. Using a module does not make a house great, however. But why 4'? I often use continuous walls of glass – full or half height – and so, using the 4-foot grid and, say, 6" square columns, a 3'-6" space is left for the windows. Many windows easily fit into that space. Or I add an extra 2x6 to each post and am left with 3'-3": perfect for a 3-foot door frame.

Four feet is a good size for closets, or two closets on the half module. Using 2' half-modules I can go to 6 feet for a bath or 8 feet for a small children's room or study. Twelve feet is good for a bed room or kitchen or small living room. Sixteen feet is another good living room size, as is the 20-foot size for a large living room or a small garage. Useful multiples go on and on.

Let me show you how I design a house on the 4-foot grid. Take, for instance, a simple rectangular floor plan like that in most "builder houses." A 24-foot width works well if all the bedrooms are to be on one side (south?)... 3 modules for the bedroom, one for the hall, and two

What module will Charles use if the nation goes metric? Will it consume fewer trees?... be more humane in scale?

more for baths, kitchen, etc. A more typical house might be 28 feet wide. The exterior walls could be located on the grid centerlines or outside, just touching the module. Partitions would fall on the grid centerlines.

Let's design a simple 2 or 3 bedroom house.

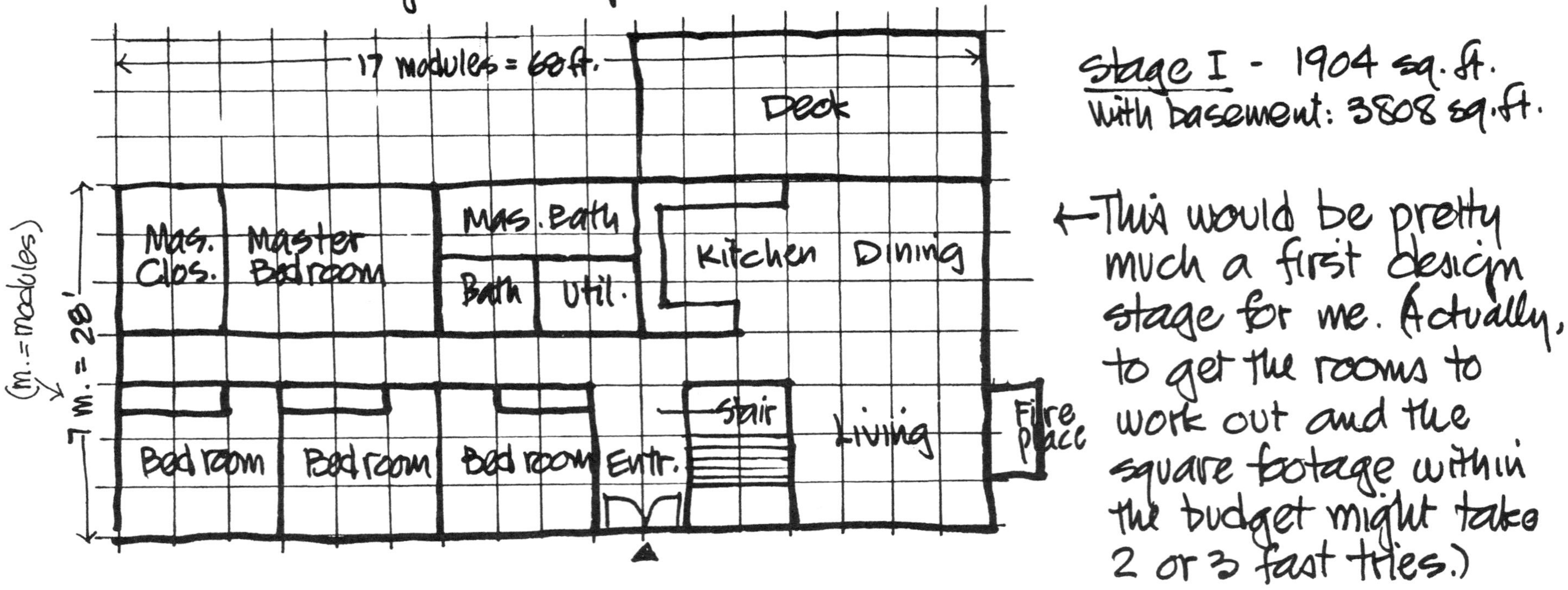

Stage I - 1904 sq. ft.
With basement: 3808 sq. ft.

← This would be pretty much a first design stage for me. (Actually, to get the rooms to work out and the square footage within the budget might take 2 or 3 fast tries.)

I find it fascinating to watch another designer at work.

Note that the closet and master bath walls are on the half module. Already, because of the grid, this is a cleaner plan than most, one that with proper windows could make a nice average house. The final plan is actually somewhat similar, not identical, to my Augello project (shown later in the book). Mike Augello had seen my work and liked it. We live in a nice but somewhat conservative town in Pennsylvania. He wanted something that at a glance would look like a typical house (i.e.- rectangular with a gable roof), a house that a local builder could build for a reasonable price. He did, however, want some brick, which is a little more expensive, but beautiful. With those limitations he said "Go nuts. Make it Woodsian!" He also wanted a four-car garage entered from the side.

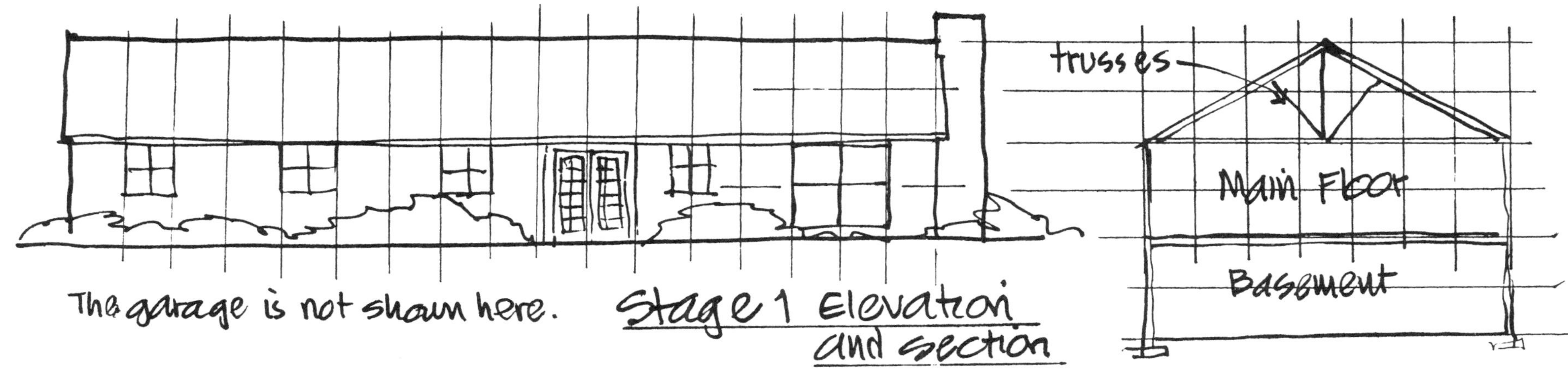

The garage is not shown here.

Stage 1 Elevation and section

Americans consume over 200 gallons of water every day.

The result was clean but boring.

So... what to do?

I decided to go with a cathedral ceiling througout most of the house, and to play off some 45° angles in the plan. They work nicely on the module. So do curves. But many designers and builders use angles haphazardly, not consistently, throughout a design. They often combine different angles (difficult to manage).

I also decided to add some wing walls and to pop out some boxes (bays) on a module or ½ module. I further decided to play with the gable roof and add a 4-foot overhang, more glass, and a low earth berm.

I'll skip some of the trial and error involved in all that in showing you the next stage. Already in stage 1 I had included a wide U-shaped stair to the basement. It's a major improvement over the usual steep, straight, 3-foot-wide stair. Now I made the deck L-shaped, added some skylights, and sunk the living room one foot, giving it a 2-way fireplace. Instead of the usual high glass I went with a diamond based on

A big vegetable garden can save hundreds of dollars in fuel and food bills. Composted food scraps enrich its soil each year.

the roof angle. To me, it's much more interesting.

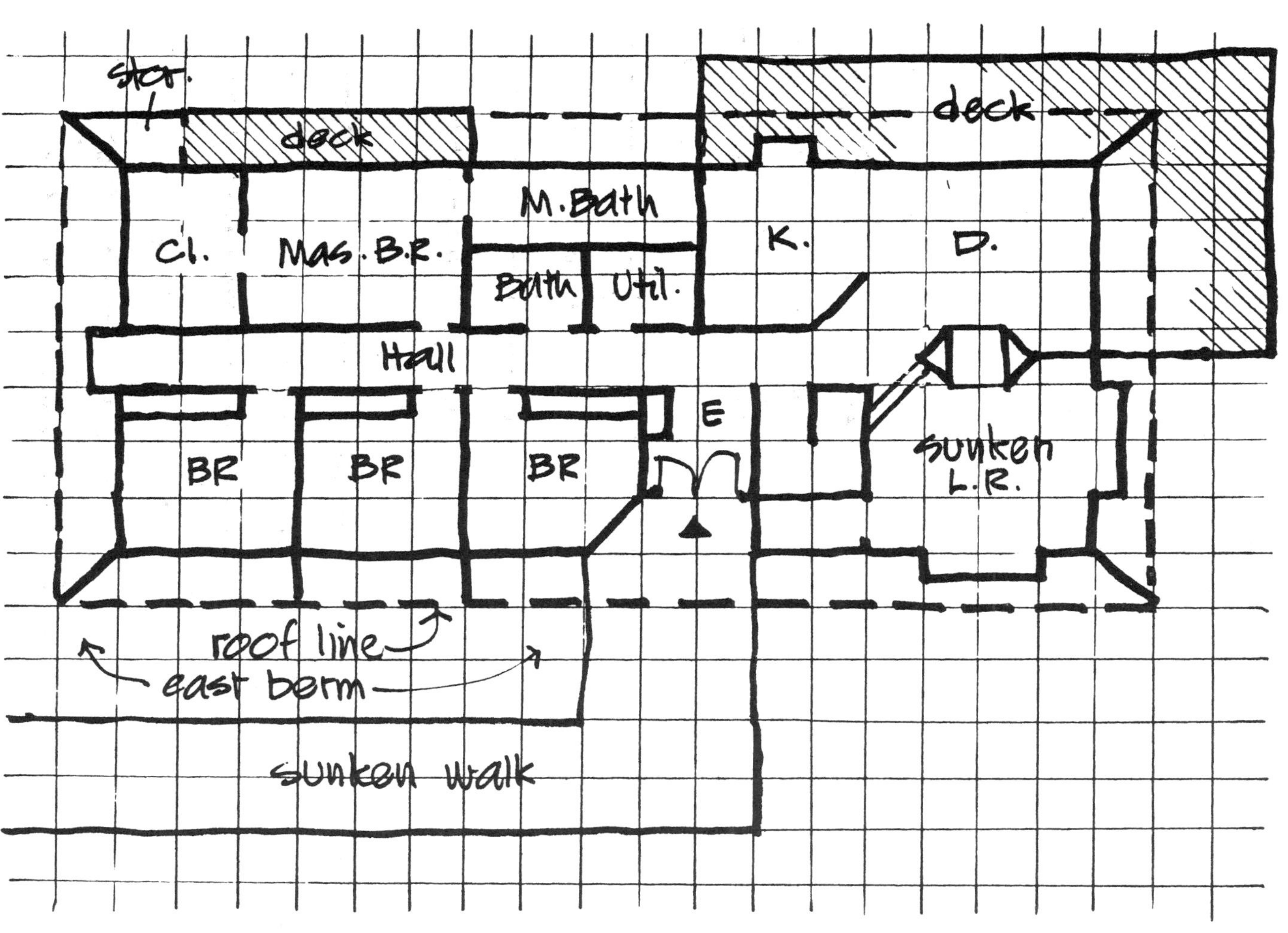

Charles has by now quadrupled the cost of the house

"Not so. It's being built for $135,000. Not too bad, eh?"

Stage II - Plan

Fruit trees, too, can reduce the number of trips to the supermarket, as can the practice of eating fewer dead cows.

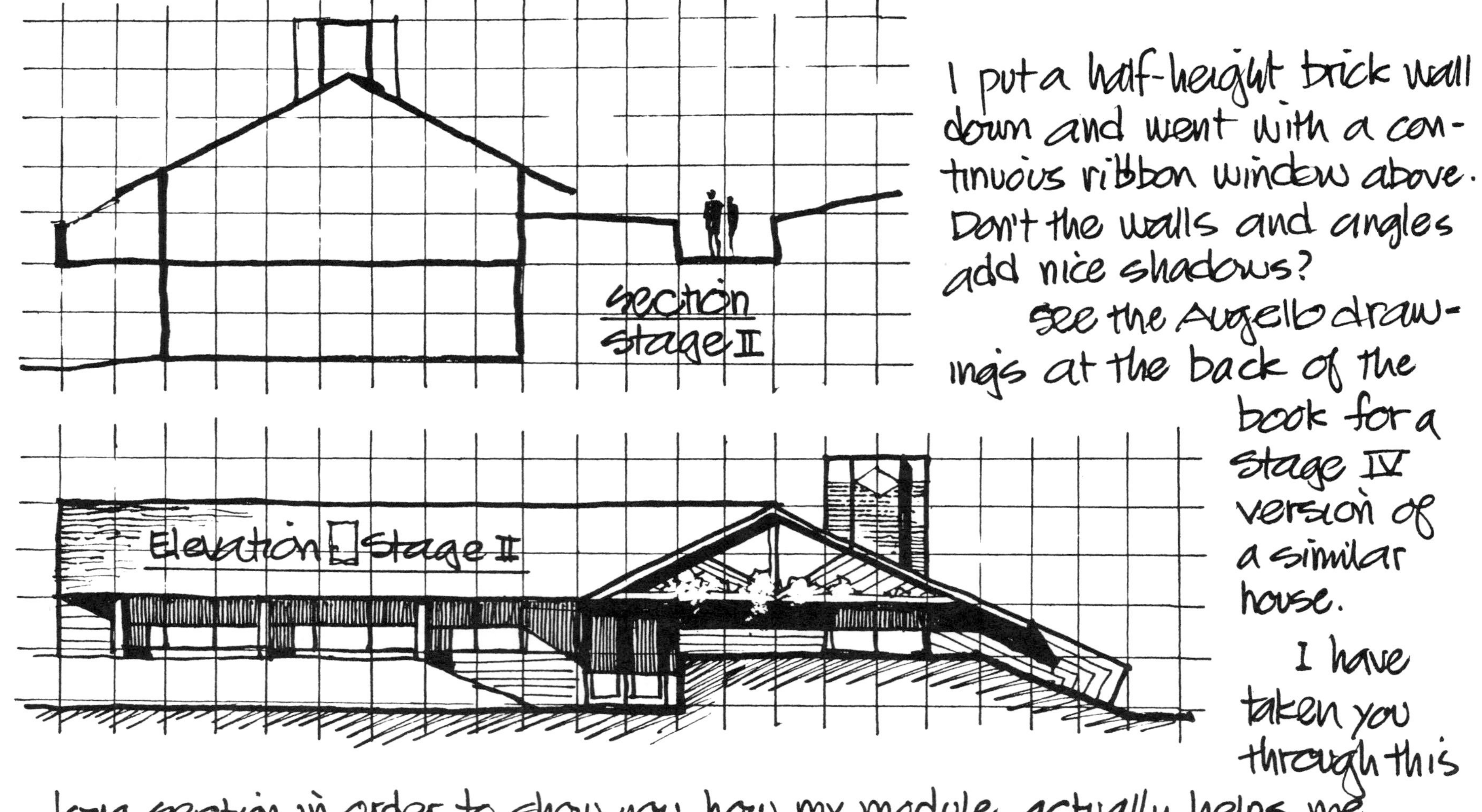

I put a half-height brick wall down and went with a continuous ribbon window above. Don't the walls and angles add nice shadows?

See the Augello drawings at the back of the book for a stage IV version of a similar house.

I have taken you through this long section in order to show you how my module actually helps me.

European-quality public transportation, hand in hand with a gentler architecture, could do wonders for America.

Sanderson Project (classic bungalow)

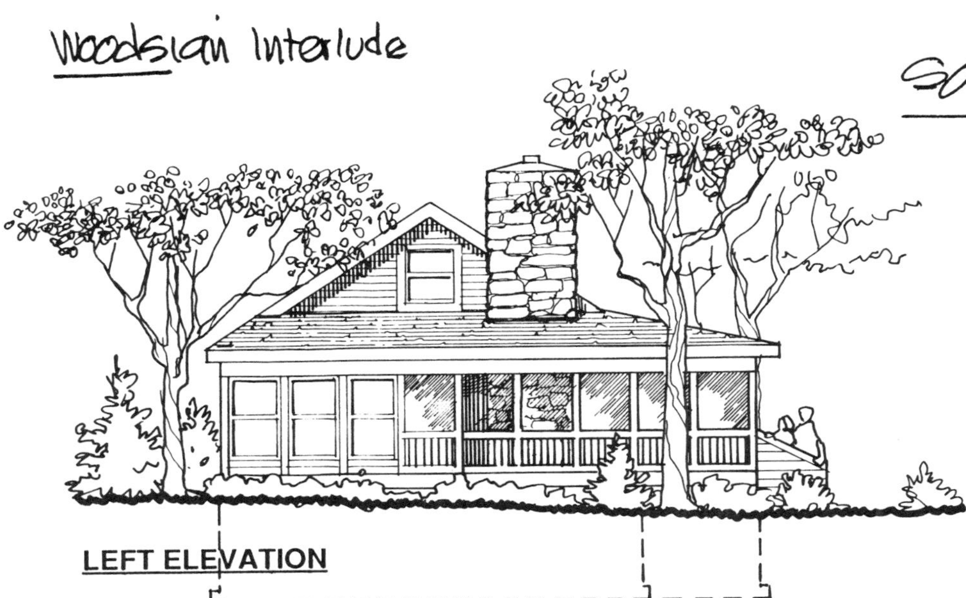

"One of my 5 traditional houses done to date. The clients thought the module was too obvious and looked too modern!" Charles adds, "I like it though. I want to show that our principles (and module) work well on traditional houses - though we don't specialize in them."

Rendering by Tracy Boyle.

The following public service announcement is brought to you as a ... well ... as a public service.

Why the following Potpourri?

Well, Charles first told me there would be about 100 points in this book. Then it was 130. And I added quite a few, too. But when the number got to 155-E(!) I decided to put all the rest into a mixed bag and call it 'Potpourri.'*

Endless ideas continued to come from Charles, and poor old Mac, who's putting all this on paper, finally had to call a halt on January 16th, 1992.

"The book stops here," said Mac, paraphrasing Harry Truman.

Anyway, even though it's all mixed together we hope it will still be both informative and helpful.

*No doubt some of the items would have warranted full pages.

If only the federal government had a renewable energy policy!

dreamer!

Potpourri, by Charles

And now Charles would like to add just a few more items... About 200!

- Siting: whenever possible, orient houses on south-facing lots so the axes are east-west. Keep the house long on this axis, if possible.

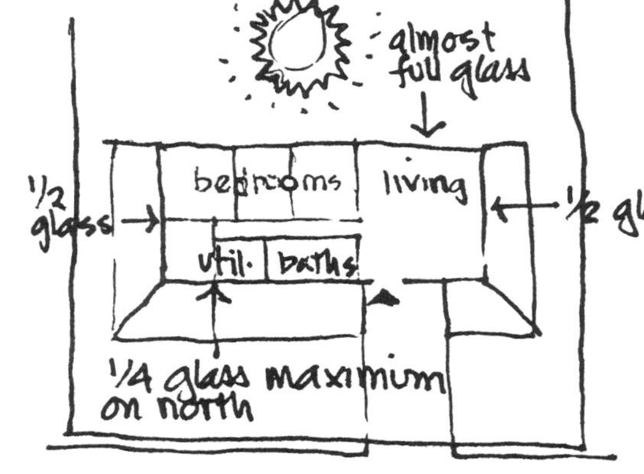

I suggest that you build an earth berm on the E, N, and W sides. Instead of having the big rooms on both sides of the corridor keep them facing south.

On the south side the overhang should be just wide enough to keep out the high-angle summer sunlight.

On the west, of course, you may need north-angled vertical louvers to keep out the blazing afternoon sunlight. On the east and north the overhangs can be sized for appearance and open-window protection.

Charles says: for consistency I often keep all sides 4'-0".
Even one large window on the north side can increase heating loads 10-25%
Keep large skylights facing south.
I highly recommend using window quilts on all windows, if possible, even those facing south.

Charles's uniform overhangs are easier to build. Mac's admit more wintertime sunlight. You've got to weigh these things.

Potpourri (cont.)

• **The greenhouse effect.** gas, oil, coal, wood ... they all dump CO^2 into the atmosphere when burned. That's why, even though we show fireplaces and woodstoves we mean for them to be used as supplements to the only safe kind of heat: solar. For romance or hot cocoa you can't beat an open fire or a woodstove but the consequences are accumulating worldwide, led by our friend, the private automobile. Even electric heat isn't safe unless it is generated by hydro power. Nuclear power is an energy horror for which we still have found no way to dispose of radioactive wastes.

• **Wood lath enclosures.** Many house owners use this material to enclose pole foundations and to screen the under sides of stairs. "In my opinion", says Charles, "it is ugly: Either cover the space with siding, etc., or use it in the horizontal or vertical position at least."

too Bavarian

preferred.

better

Charles's choice

• **Garage door structure.** Beams (see next page.)

Gets complicated, architecture, doesn't it? It's really a moral issue.

Potpourri (cont.)

above garage doors often sag. To prevent this use a flitch plate* (steel sandwiched between two-by^s) or a steel beam. Where to put the entry door in a garage is often a problem. The best place, at times, is between the overhead doors but often a supporting column is there. This is what I have done:

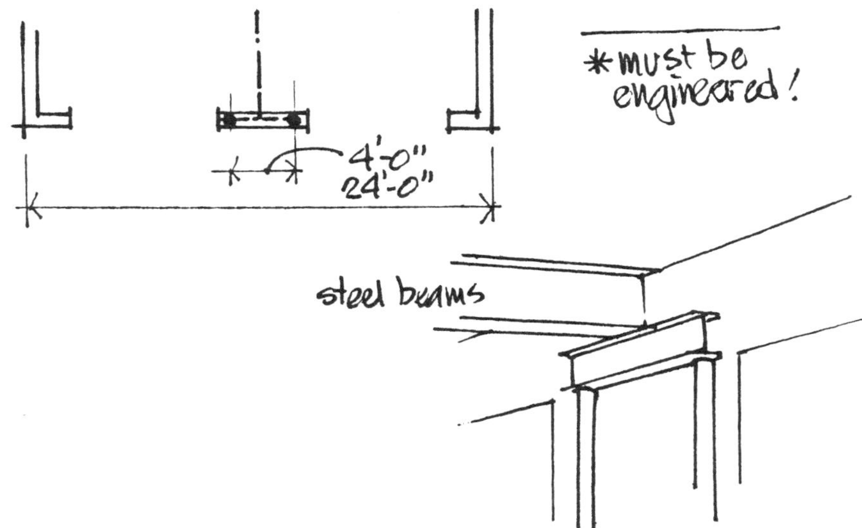

*must be engineered!

And here, Charles reminds us that there are other possible door positions

• He further reminds us that he likes to conceal electric meters. Here's a plan of one tucked into a wing-wall at a garage.

If, as is more and more the case these days, the meter must be located near the street, it can be housed in something that relates it to the house design.

Here we take a short sabbatical from moral values.

Potpourri (con't) - still by Charles

- consider using cedar and better* siding if possible. In the long run it will save $, and in the short run will look great.

- Angled windows not at the angle of the roof look real bad! or - if cutting off so little at the top leave a point. Preferred:

- some energy-efficient house types that have appeared over the past 15 to 20 years are the following:

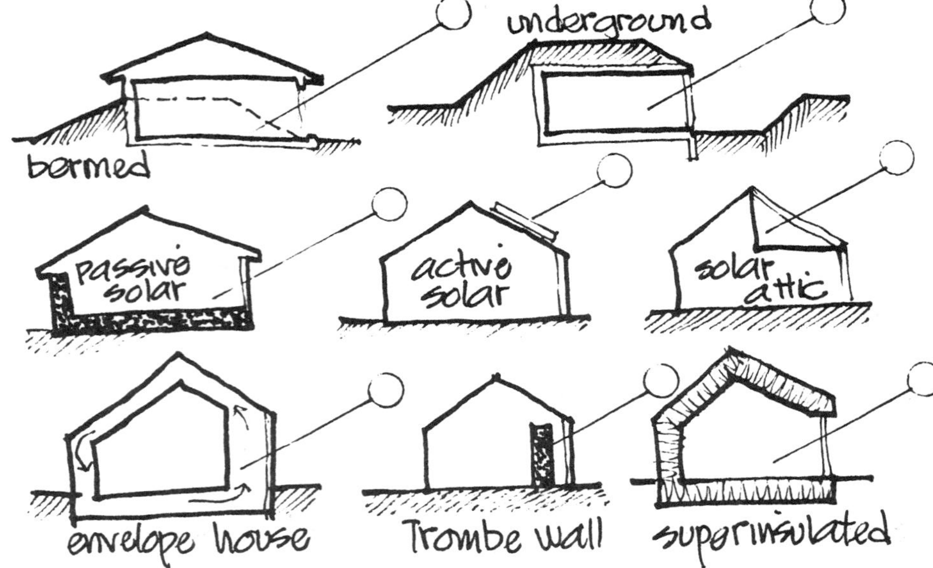

Many of the types can be synthesized. Most of my houses are passive solar, superinsulated, and earth sheltered.

- consider 2 shower heads in the shower stall for 2-person showers - we've all done it at least once!

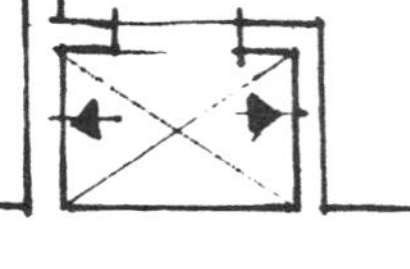

---

* I hope he's not going to say redwood.

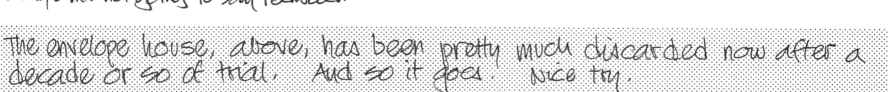

The envelope house, above, has been pretty much discarded now after a decade or so of trial. And so it goes. Nice try.

Potpourri de Charles (cont.)

- On large baths give toilets privacy.

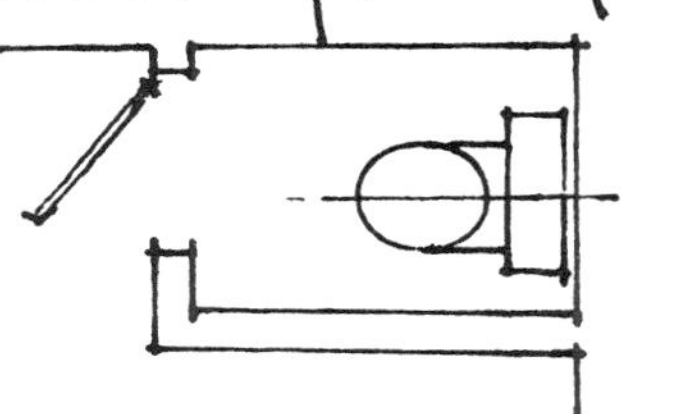

- Medicine cabinets on sides of mirror that runs full width of bath cabinets.

- Use dimmer switches on almost all lights. It costs a little more but gives you cozy lighting.

- Most basement stair doors must swing out. If possible, add a safe top landing and let the door swing in on it.

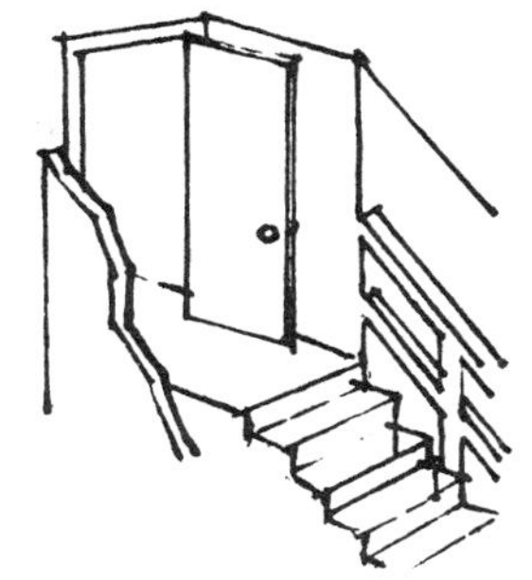

- Types of earth homes:

They're where the earth lives, aren't they?

bermed

earth sheltered

underground

underground.

I do not recommend this. Rear rooms get too dark, and skylights do not provide the required second means of egress. I suggest that you bring the earth up onto the roof in a couple of places, and in the rest do this:

Just don't go underground for the wrong reasons. If you do it only to impress the neighbors you're bound to overlook something important.

Potpourri "Charles"

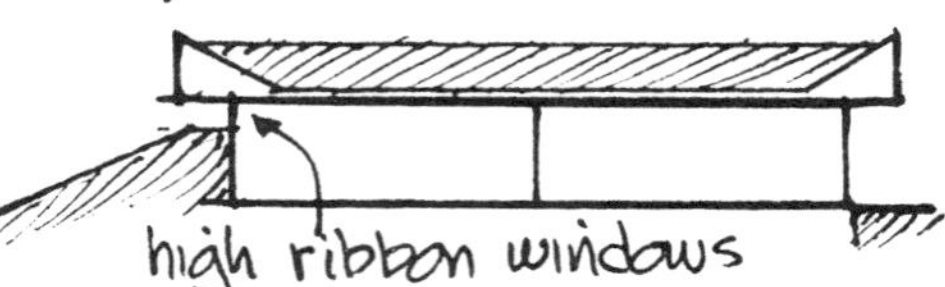

or use an atrium design:

In any case you must provide fences or other barriers to prevent falls from the sudden drop-offs. "Parapets, Mac! Ha ha."

- Use sunken, raised, or built-in conversation seating.

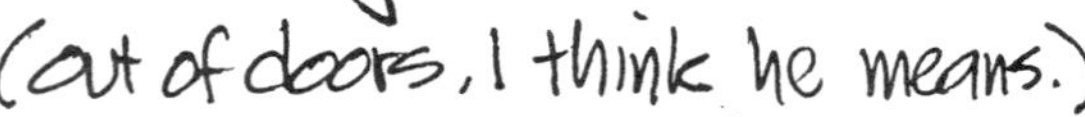

On second thought, I guess he means indoors, too, so here's an indoor version:

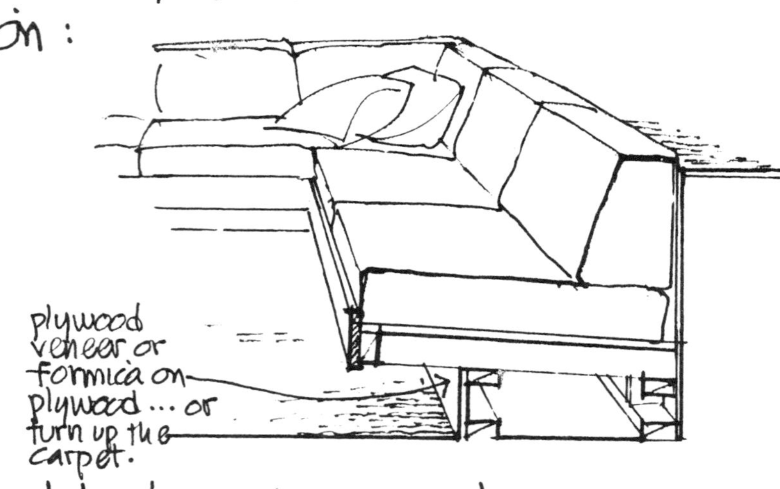

plywood veneer or formica on plywood ... or turn up the carpet.

- Glass block walls can be beautiful. (Remember, that's Charles speaking.) "Yes they can, Mac."

- I love standing-seam, Terne metal or copper roofs but beware: very expensive.

- If you can afford it put glass doors on bookcases and most open kitchen cabinets.

Be careful of atrium designs if you want to have passive solar heating.

Potpourri eternal (cont.)

- Houses always look too high to me.

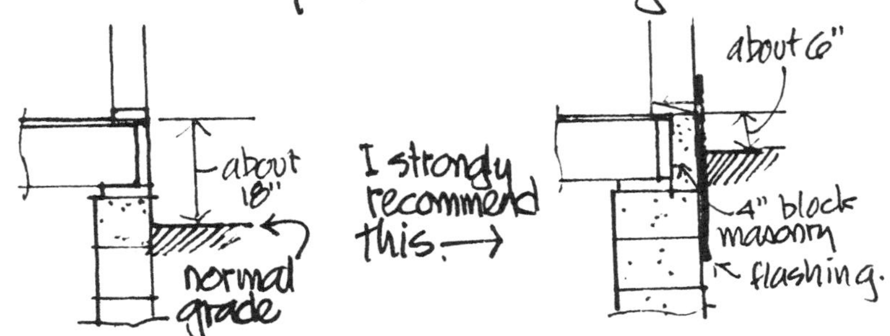

- High berm. I use this on the rear of garages at times.

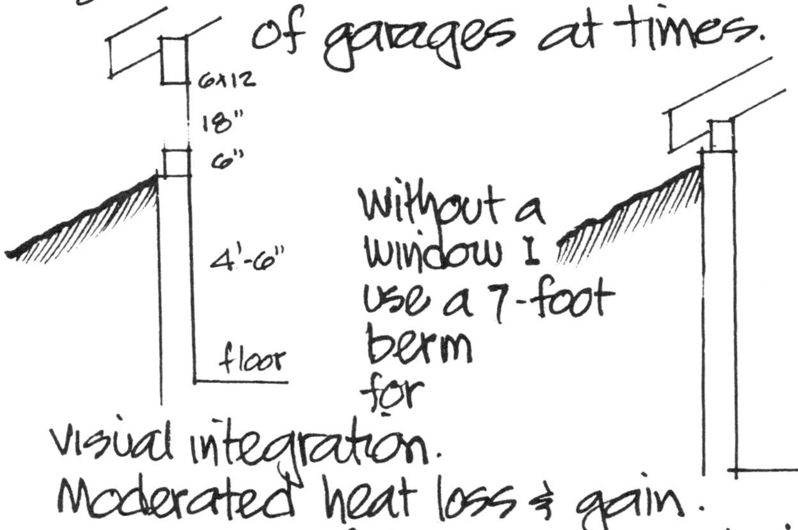

Without a window I use a 7-foot berm for visual integration. Moderated heat loss & gain. I use both of these wall sections on the north sides of houses (bath rooms, utility rooms, etc.)

- And here Charles shows us how he details his earth-covered roofs.

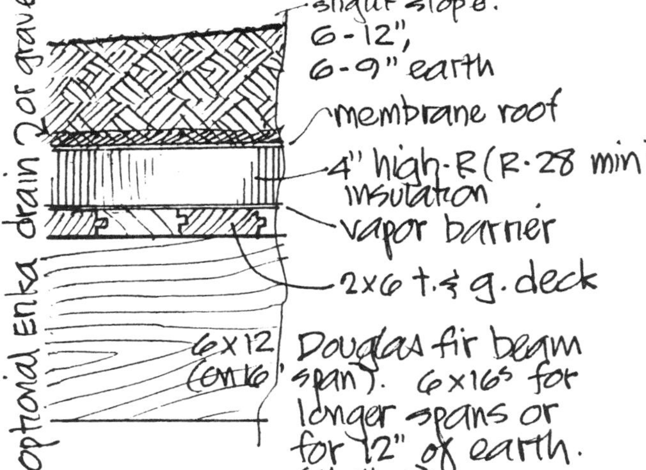

I would like the record to show that I take issue with this approach. *Putting the main water-barrier membrane atop the soft insulation board seems to invite puncture. I should also point out that 6" to 9" of earth requires summerlong irrigation if you don't want a dusty-dry landscape. Roofscapes should start with a stony or sandy drainage layer, some fill dirt, then a rich layer of topsoil covered with rotted mulch - humus.

"I've done it my way a lot with no problems," says Charles. "But I also prefer 12" minimum earth with gravel walk."

Read all you can about earth-cover techniques, and talk to people who've lived underground, before you make your move.

The Potpourri goes on...

- Use walk-in closets where possible ...with a window if possible, or a skylight.
- Stain cedar roof shingles at times.

  between 8 and 12:00 is probably best

- Use double interior doors where appropriate.

  Master bed room and family room.

- Use trellisies.

  use copper pipe from demolished buildings makes a lasting and handsome top rack.

  The vines drop their leaves just when you need sunlight.

- Use good pocket doors where there are space problems.

  (use bad ones where there aren't?)

- Consider using post, plank, and beam construction.

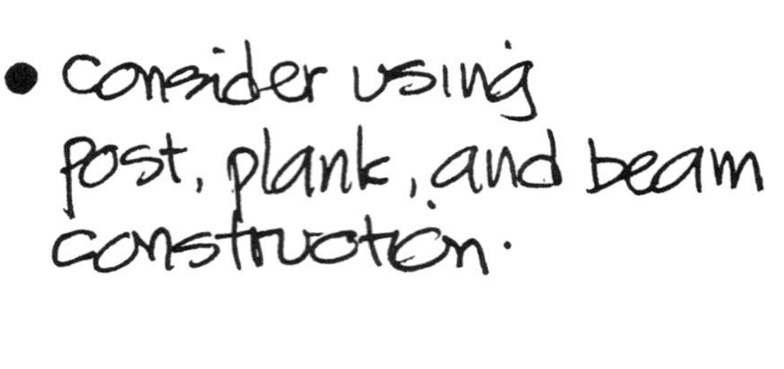

- No storm windows on new construction!

- The Lennox Pulse Furnace is up to 97% efficient. "Do you mean to say," I asked Charles, "that only 3% of the heat is wasted up the chimney?" "That's what the specs say," he responded. Impressive.

  Hmm...

- Use one-piece sink/counters or sinks mounted to the bottoms of

I hope you took a sturdy canvas shopping bag to the supermarket this week.

Potpourri forever . . .

counters.  Top-mounted sinks create countertop puddles by preventing run-back into the sinks.

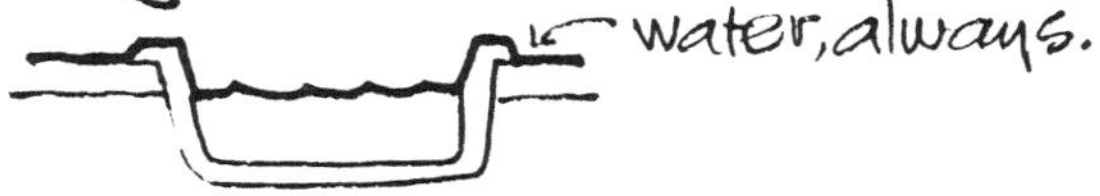

water, always.

- I like heating systems with hot water running in butyl rubber pipes through the floor slab.

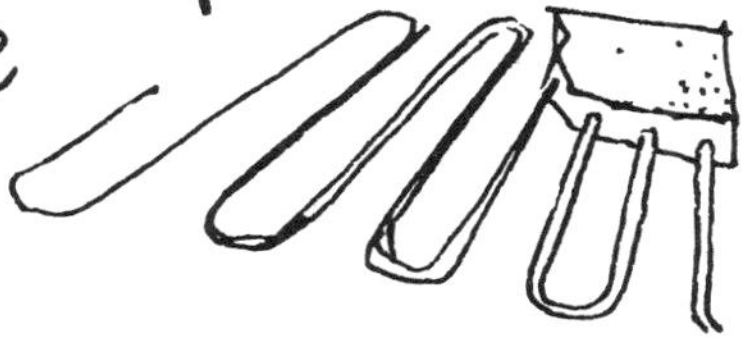

- "Everyone," says Charles, living in gray stone country, "uses gray stone on fireplaces, which is OK, but on earth-tone buildings use earth-tone stone (brown, gold, reddish), and dye the mortar.

- If you want to build an earth-sheltered house, buy my books. Don't build a bomb shelter. Admit daylight to every room, and use a heavy timber roof structure. "Buy Mac's books first!" adds Charles.

- Use breezeways and screened porches on the same module as that used for the rest of the house.

- Use water-resistant plasterboard in bathrooms, greenhouses, and kitchens.

- Split-face concrete blocks are great beautiful products, and

It's OK to take notes as you read all this. Just be sure to take them on scrap paper. And don't tear the scraps from this book!

Charles's ongoing potpourri...

would solve that exposed foundation problem.

- I like fiberglass (or equivalent) heating ducts. Quiet.

- I like carports. So did FLLW.

- If the furnace chimney is separate from, say, the fireplace chimney make sure it matches the house and is at least 2 feet square!

- Use two-story areas with lofts in the living areas of a two-story house.

- study Japanese architecture, for examples of excellent simple building. I also like Scandinavian Gothic architecture.

Charles! Charles! This is a simple little book of design. They don't have time to go study Japanese architecture! "sh-h-h!" responds Charles.

- Decks can be ugly from below. Use spaced boards of cedar or pressure-treated wood on the underside if you can afford it.

To which the old grouch feels compelled to respond, "I see bits of leaves and paper, old bird's nests, rot, and fire hazard in this." — "clean it every 10 years"

- Use clerestories, with overhangs.

←clerestory.

Whoa! Slow down, Charles. Let's see: you like glass block and fiberglass ducts and cedar siding and...

the potpourri goes on . . .

- Do not use large above-ground pools in small backyards, but if you do be sure to cover their sides with material to match the house. Use in-ground pools if possible.

Many children drown in backyard pools each year. Comply fully with all safety codes ... and watch those kids!

- Use small reflection pools of concrete block or pressure-treated 6x6s with butyl liners, weep holes, and underwater lights.

And follow the same safety precautions as in the item above!

- On steep sites consider using terraced siting. Steep sites are vulnerable and must be protected with great care.

This looks like a fairly simple proposition until you think about the hundreds of cubic yards of earth you must remove to create safe working conditions at foundation level. All that earth must be removed by a convoy of trucks. Later, much of it must be trucked back and replaced, layer by layer, as it is carefully tamped against the walls. — M.W.

Be sure to engage an architect or an engineer if you get into this sort of thing. (still great for steep sites! - CGW)

- I love spiral stairs ... in some places up to 8'-0" in diameter. Can pass code for main stair, says CGW.

swimming pools are often filled with water.

Potpourri, potpourri...

- My favorite siding is 1×8 channel groove cedar siding. It gives strong horizontal lines and shadows.

rough sawn

- Do not use metal closet doors unless they are of good quality.

- Slightly sunken driveways can really help make houses look lower with less basement exposure.

pressure-treated 2×6s

- When the garage is at the lower level, *why* make everyone climb out of doors to the entrance? An inside stair right there at the garage, out of the weather, often makes good sense.

garage
house

- "An open, flat site", continues Charles, "wants a *horizontal* design or a very *vertical* design, like a tree."

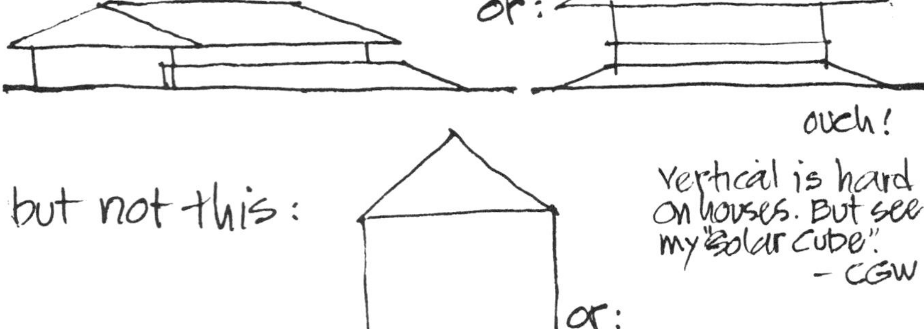
or:
ouch!
but not this:
or:
Vertical is hard on houses. But see my "Solar Cube".
– CGW

We printed this book on recycled nuclear wastes just so you wouldn't have to feel guilty about participating in the use of trees for the project.

Potpourri, still by Charles.

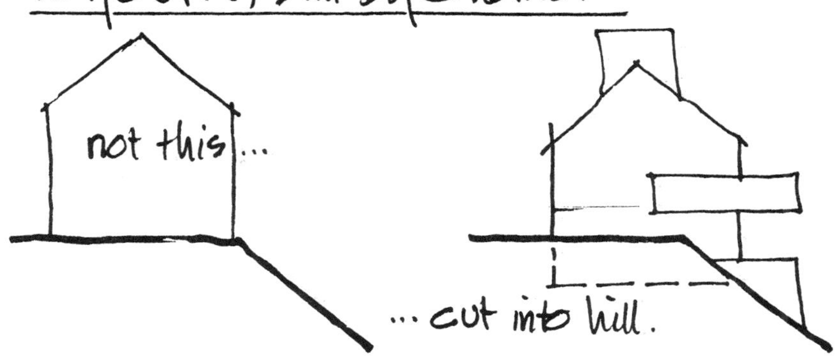

- As houses get higher I generally dislike them more.

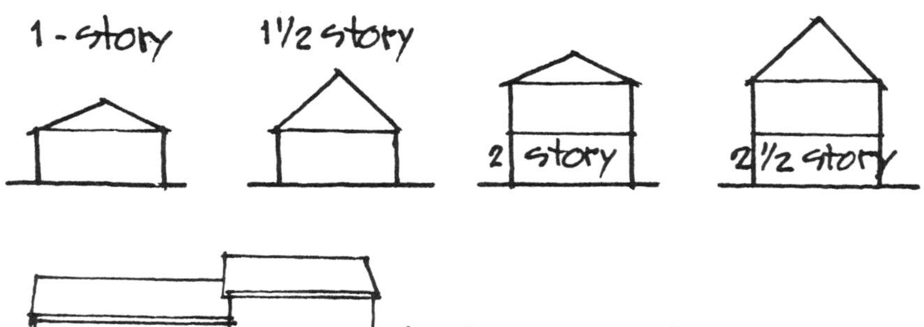

bi-level can be nice.

And I say you forgot to include the ½-story house, Charles.

(Mac's really mean, people.)

- Charles likes the "reveal" in some buttresses I've designed. I got the effect by lining the top of the form with sheets of ¼" plywood. Also, I did no vibrating of the concrete in order to get the rough, honeycombed look of raw concrete.

- "Less is more." – Architect Mies Van Der Rohe

"Form and function are one." – Architect Louis Sullivan

"Cut doors and windows for a room. It is the holes which make it useful." – Lao Tzu.

Just think: by now you, dear reader, are a well-informed designer and we haven't even reached the end of the book.

Pot... pot... potpourri

- Frank Lloyd Wright overhang:

- I recommend casement windows. They offer a 100% opening. Double-hung windows* offer at most 50%

- Mac, say something on air conditioning.

B-r-r-r-r.

OK, I'm sorry. Air conditioning is a delicious phenomenon on a blistering summer day but it brings with it a great penalty. Several, in fact: gross air pollution from the overworked power plants, gross energy consumption. It has also helped remove us further from the natural world and given us the sealed, faceless towers that now characterize every city on earth.

Estimates vary but it is generally conceded that 50% of all air conditioning could be eliminated if our buildings were properly designed. My little underground gallery would need no cooling if the customers would stay away but they let in so much humid air we must now

* Frank Lloyd Wright called them guillotine windows.

It must be noted that the human body has its own automatic cooling systems that work quite well if too much is not asked of them.

Potpourri·i·i·:·:

run a tiny cooling unit occasionally to do the needed dehumidification.
Shade trees, earth covered or reflective roofs, proper orientation and overhangs, proper attic ventilation and insulation... such things can make enormous reductions in your cooling loads, as will thoughtful summertime cooling, laundering, and bathing activities.

- The built-in gutter. ("important detail" - CGW)

*see note at right.

cut 2x12

shingles on plywood

drip edge

seamless gutter! (sloped)

drop (no ice dams!)

2x10

2x6 treated

rubber flashing covers back of fascia.

wing wall

continuous screened vent lets condensate drip out.

siding on plywood

downspout concealed within wing wall

- A 1½" ø pipe railing, painted earthtone, is nice for railings that are not solid. Charles goes on to ask if I remembered to emphasize somewhere his liking for solid railings, or at least partial ones, and to tell you the truth I can't remember if I remembered or not. sorry, Charles.

* "This is a really important detail, which Mac should not have put in 'Potpourri'! He knew it and he did it anyway!" "Did not!" "Did so!" (Detailed by Larry Wilson - but check for local condition problems.)

Hang on, folks; I think I see the end of the book not far ahead.

Potpourri

- When it comes to rooftop solar heaters Charles says, "I hate this."

And I'd like to add that I hate this.

Mount them on the ground, adds Charles, or at the → same angle as the roof

In any case we urge you to use a solar domestic water heater called the Copper Cricket.* It gets rave reviews from all the top environmentalists and energy experts.

- Corner windows can help open up a usually closed part of the window design.

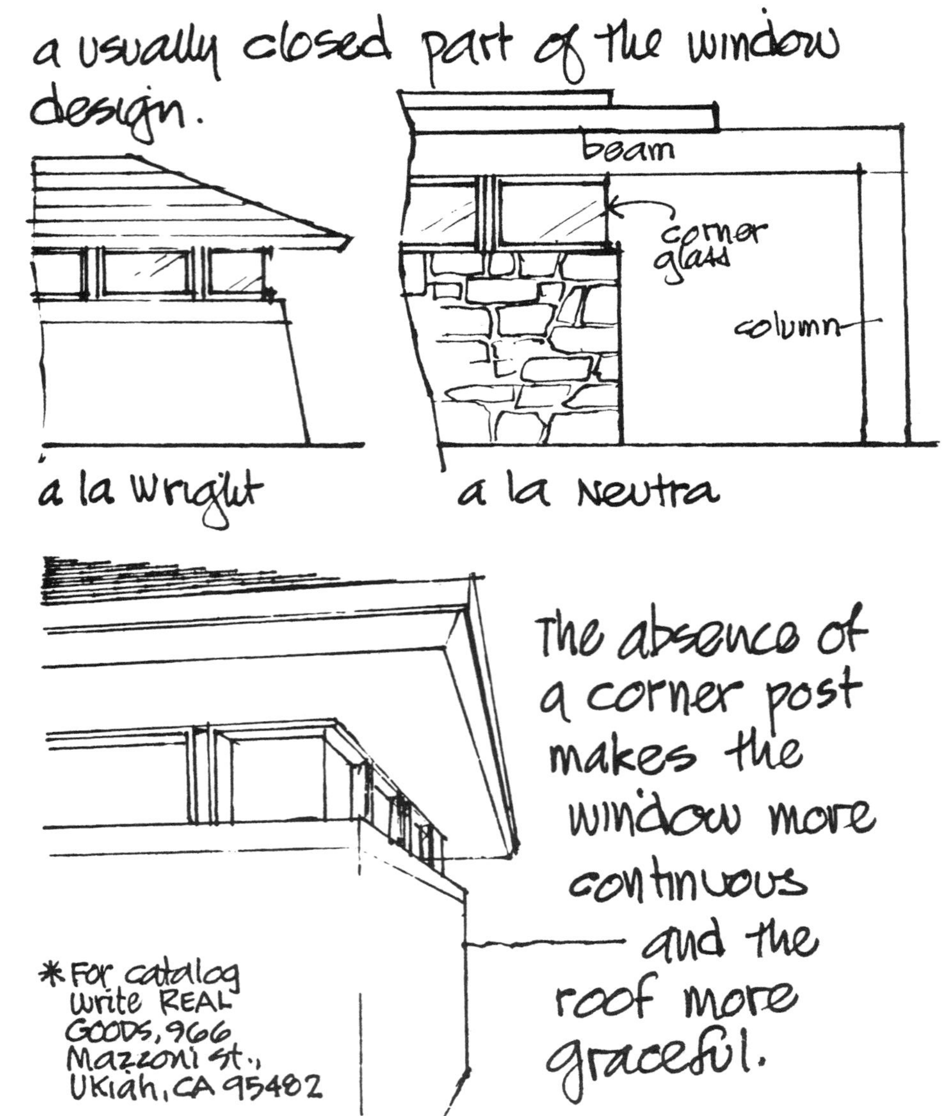

The absence of a corner post makes the window more continuous and the roof more graceful.

* For catalog write REAL GOODS, 966 Mazzoni St., Ukiah, CA 95482

Domestic water heating is one of the greatest energy users – and wasters – in America

Potpourri now, potpourri forever

- Basements are not needed. If a house is bermed it is less expensive not to have one. (A basement, I believe he means.) You can have a 12' × 16' storage room attached to the garage. Also lose the attic. Get rid of the junk.

- Consider putting the laundry room near the bedrooms instead of the kitchen.

- Flat roofs are OK, and will not leak with proper butyl rubber roofing. Flat roofs and shed roofs are the least expensive. Let's go back to them.

- Instead of miscellaneous TV, stereo, and computer equipment combine them into one wall system made of 2×4s, plywood and Formica.

- Put skylights in interior baths and large closets.

- Use built-in beds in small children's rooms.

(Remember: no built-ins in large children's rooms!)

EMR — electromagnetic radiation — in the home may be harmful. From TVs and computers to electric clocks and electric blankets the zapping goes on.

Potpourri (which, as you'll recall, is brought to you by Charles)

- Because of the high costs of masonry fireplaces I have come to like zero-clearance fireplaces. Wrap them with 2x6 studs and insulation and you can cover it with anything... siding or brick or stone veneer. This saves about half over full masonry fireplaces! These prefabs even burn off gas.

- I suggest that you use 1-piece tubs and showers when money is an issue. Job-constructed showers and raised tubs can be beautiful, though!

- Pressure treated (ground contact) 6x6s make good retaining walls. Connect with stainless steel pipe.

steel pipe

4' fork

gravel

- Wherever space is tight use good quality sliding doors at baths, closets, etc.

- If you build a deck above a living space use butyl rubber flooring, floor drains, and removable cedar decking and scuppers.

- Soundproof walls by using staggered studs in baths and bed rooms. The are also metal Z channels you can use.

Hang onto your values, folks, as we continue our walk through this supermarket of building materials and equipment.

- When using cathedral ceilings remember you need collar beams, a ridge beam, or massive walls.

Without one of these, the roof will collapse! *

- My favorite woods (says Woods) are for construction framing: Douglas fir; for siding: cedar (R.S.), for some trim, flooring, and stairs: oak.

- No curved trim inside on modern houses.

- Always conceal or bury gas or oil tanks

- Use water-saving faucets, showers, toilets, etc.

- Structural tricks I use a lot. Be sure to use an engineer for your job!

Doug. fir

5/8" plywood between 2x12s covered with siding

For long spans in cathedral ceilings!

"Glulams are probably less expensive, I guess." — CGW

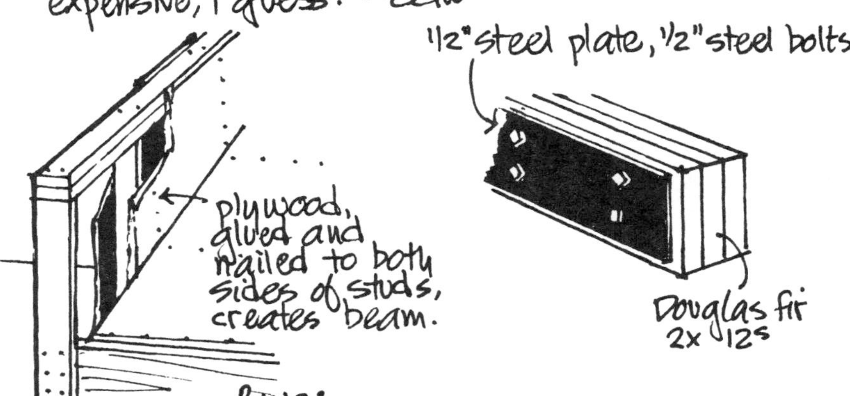
1/2" steel plate, 1/2" steel bolts
plywood, glued and nailed to both sides of studs, creates beam.
Douglas fir 2x12s
2x12s
Look! No beam below

* Don't let the roof get too thin. It needs deep insulation as well as a clear airway above the insulation if it is to perform properly.

Potpourri

- Charles goes on to ask himself, "Will these suggestions cost more?" To which he responds, "Some no doubt would, but some would save you money!" And he continues...

  Good proportions, consistent use of windows, style, and trim cost nothing. My basic houses, throwing in storage and garages, cost a minimum of $40 per sq. foot. My average houses are costing $60 per sq. ft. My most expensive houses to date are going for $125 per sq. ft. I could of course spend more. Some custom houses cost $300-400/sq. ft. Even using all these suggestions my typical houses are about average. So enjoy! Good plans, and working with the builders on bids help a lot however. And supervision can solve a lot of mistakes.

- If you must build in the style of French Provincial, Southern Colonial, English Tudor, do it right or not at all. It will take research and cost $.

- For half the clients in the world who ask: "log cabins are not cheap!"

  This is so deep I cannot imagine what it means.

- Use care in interior decorating — pictures, etc. consistently.

  Ditto.

Cost is not the criterion. Nuclear power plants cost more than any other structure. That's OK. But they're fiercely dangerous.

Potpourri

- I strongly dislike winders in stairs. If you must use them, comply with you local building code.

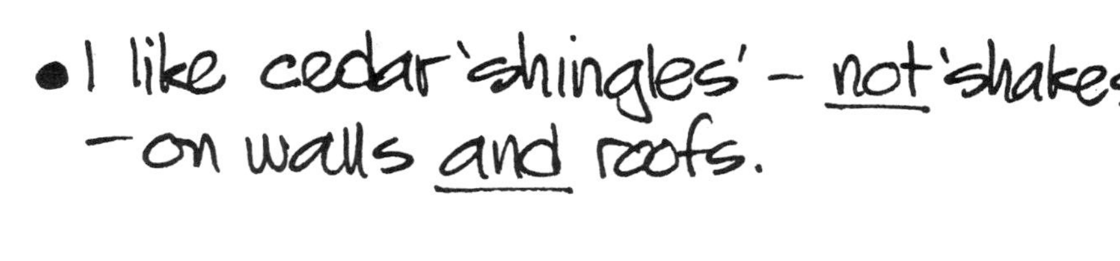

- carefully integrate heating ducts.

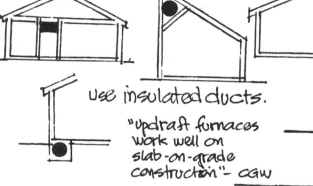

use insulated ducts.

"updraft furnaces work well on slab-on-grade construction" – CGW

flashing

- Bring brick veneer down to grade.

- I like cedar 'shingles' – <u>not</u> 'shakes' – on walls <u>and</u> roofs.

- Use earth-tone kitchen and bath fixtures – ivory, tan, even rose, sand – stay away from blues, pinks, etc.

- Windpower and solar electric (photovoltaic) devices can cut you free of the power company. see the Real Goods catalog, free from 966 Mazzoni st., Ukiah, CA 95482.

- consider adding a strip of another color as a carpet border. same material. This need not be expensive.

darker terra cotta, etc.

Of the world's top ten fuel-efficient cars not one is an American brand.

Potpourri (only 100 more to go, folks)...

- More on garages: I highly recommend detached garages where cost is a consideration. (Charles no doubt means the <u>use</u> of detached garages. Let us continue:) They can be uninsulated and uninsured. Your policy should insure out-buildings for 10% of policy. You can connect garages by breezeways. I would also recommend berming garages. They will not require deep frost-footings and the bermed earth will modify temperatures a bit inside. I know that many people build re-slabs on gravel, but I am not sure they will last for over 50 years.

what's a re-slab?

"Reinforced slab".

- I also suggest (the use of) surface bonding on concrete block.* It's easy to use, strong and low cost unless you lay it up conventionally with mortar as many masons want to do. Then it's expensive and not as strong.

- Use deciduous trees on the south and west sides. They let sunlight in in the winter. Use evergreen trees on the north side.

cold wind

- For traditional houses keep mostly symmetrical, for modern use more carefully balanced asymmetry.

"Are you sure, Mac? I've used it successfully."

*careful! surface-bonded block often fails under earth pressures. Even a sharp bump by a bulldozer will crack its structural skin!

Potpourri

- I think on traditional houses that double-hung windows look best tall rather than wide as is often shown on recent designs.

Casement windows should be square at widest, long look good too.

Sliding windows look great in wide proportions as do awning units.

But I don't like deep awning units.

Some window companies have at times offered sliding glass doors cut down to different heights for windows. When it is not cost-prohibitive it is one of my favorite window types.

All doors and windows can be the same width.

Where different-width windows are needed make them proportional. For instance, if your basic module is 4' make smaller one two feet.

They look as if they go together.

- Low cost housing: is it possible? Yes. At least lower cost. I would use bermed slab construction with a shed roof. Or, if more area

As heat is lost through windows it is unaware of their proportions.

Potpourri (cont.)

were required, I would still use a bermed structure with rooms upstairs under a steep 12:12 roof. Both could be built economically. I would keep garages and storage buildings separate, and stick to a simple rectangular plan.

I would keep sleeping rooms small and use built-in furniture. I'd make the house passive solar, keep the finishes simple... dyed cement floor, surface-bonded walls, etc.

If someone acted as his own contractor and did some of the finishes he could probably save half... build a $100,000 house for $50,000!*

"Important note!" - CGW.

- What if you live in an ugly house?

I want to be clear: I believe that beauty and goodness are related. In God they are identical, in persons, interrelated. But many kind people live in what I would call ugly houses, and many evil people live in beautiful houses! Our culture is all askew! 200 years ago almost anyone could build a beautiful, simple dwelling. Now almost no one can, not even most architects, not really beautiful and simple. I include myself in this ultimately. Why? That would take another book. My point is that saving the ozone layer or feeding a less fortunate brother or sister is more important than whether we have plastic columns. But after saving the planet and morality, Beauty is important!

*Similar to Mac's Popular Science house at the back of the book.

I heartily endorse the column on the right. This one. →

Potpourri, P.S. Some afterthoughts from Charles Woods.

- Do not use different drapes on modern houses. (Different from what?) It looks dumb on the outside of the house.

- Be very careful in the use of ornamental iron railings. You don't want to fall and hurt yourself. Is that what he means? Or does he mean that one of these is a little too clever for its own good?

- On roof undersides I prefer angled versus flat. Also: use siding, where consistent, on that surface and not plywood. Or, if plywood is used make sure it's of top quality.

I'm not saying this is bad.

Only this is usually better.

- Frank Lloyd Wright often used stock concrete block for exposed walls. He cleaned and sealed them and raked out the horizontal joints. It still makes a nice inexpensive wall.

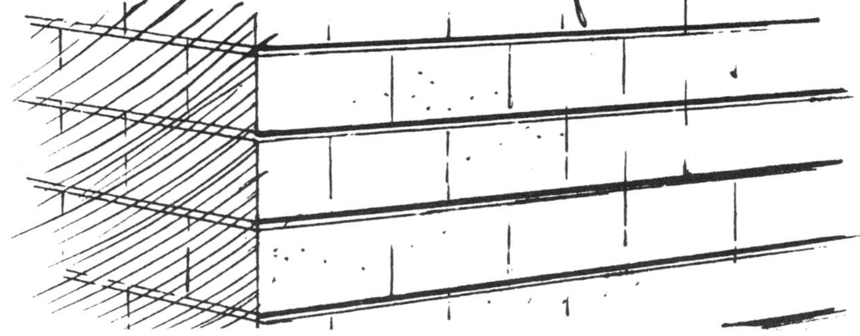

Today we're all going to sit down and write letters to our congressmen and -women. The message: There's trouble on the land. Don't you think they should know about it?

Son of Potpourri

- On driveways, use red or brown gravel over gray. I hate seas of gray gravel!

- Maintenance. People often ask for no-maintenance houses. There are no such things, but clad windows are great, and stone, brick, and block don't require much upkeep.

  clad = factory-applied vinyl or metal covering of all parts exposed to the weather.

  Stain holds up better than paint which fades and chips, while even weathered stains on wood like cedar look great.* Flashings should be of anodized aluminum or copper. (I do not like aluminum or vinyl siding but it might be a prejudice. It holds up pretty well.) I like cedar shingles (not the rough, handsplit shakes) for walls, too.

- Rainwater conservation: if water is piped away to the street, the storm sewer, and the nearest stream it does no good to the land on which it falls. It also causes flash flooding and erosion down the street near the Jones's place. If your soil is porous create a sunken garden/percolation bed to hold the water till it can soak in. If your soil is impermeable use the sunken garden as a retention basin and let the water run off slowly through a small drain. Perhaps we will show you these on the next page.

* Be careful regarding materials meeting. steel nails break down aluminum siding, and cedar breaks down copper!

Remember to dress warmly indoors in the winter. Let your internal heating system do much of the job. Don't make the furnace do it all.

Son of Potpourri (cont.)

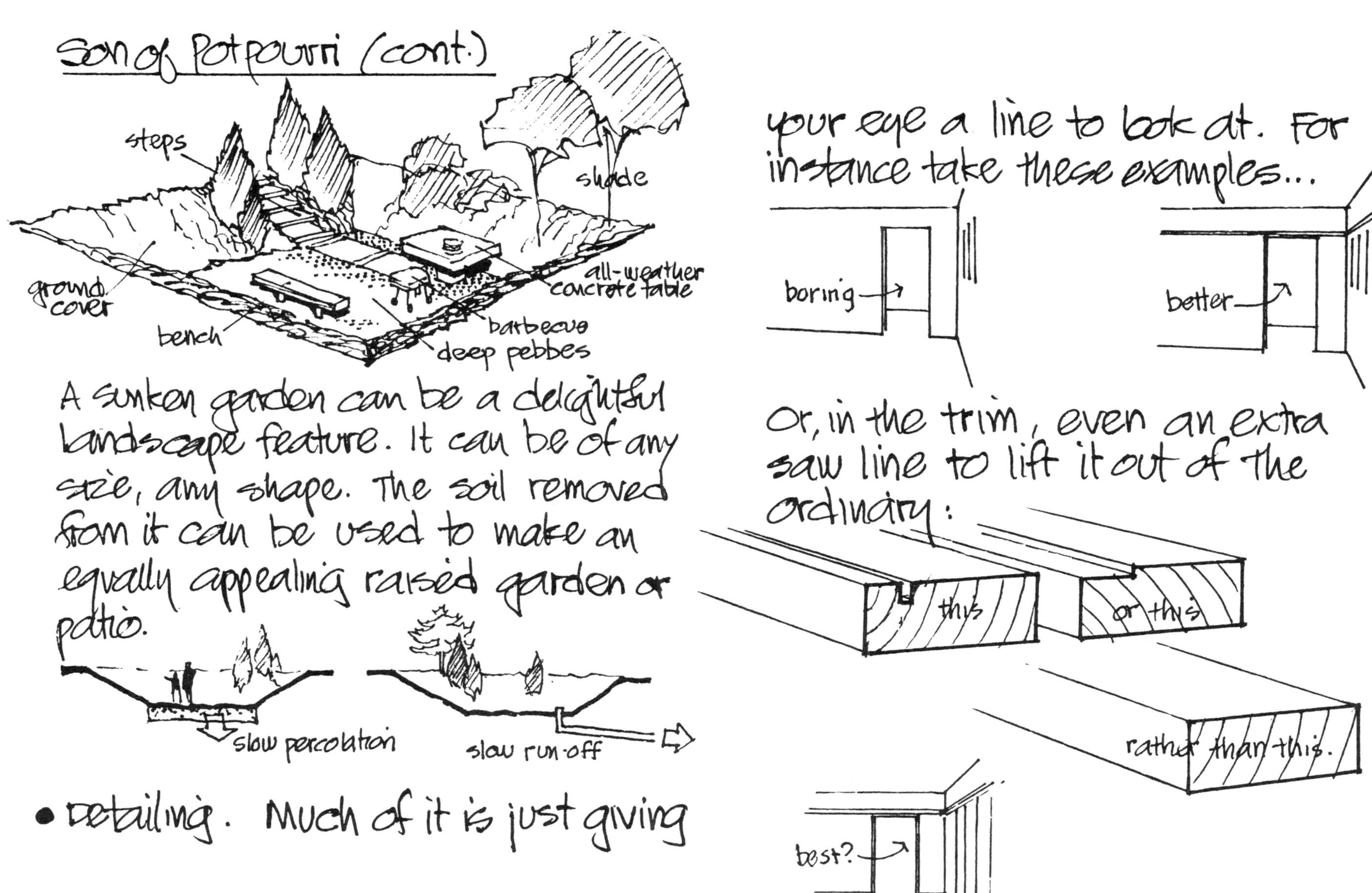

A sunken garden can be a delightful landscape feature. It can be of any size, any shape. The soil removed from it can be used to make an equally appealing raised garden or patio.

- Detailing. Much of it is just giving your eye a line to look at. For instance take these examples...

Or, in the trim, even an extra saw line to lift it out of the ordinary:

If you see a sunken garden, chances are you'll find a caring person nearby.

Son of Potpourri, finale.

• Frank Lloyd Wright often said he wanted to "destroy the box." That's why he used open plans, corner windows, etc. In my own work I have tried to limit the divisions between floors, walls, and roofs.

glass

sloping walls

That's why I use my wing walls and earth berms, too...

I've always wanted to do a circular house like this:

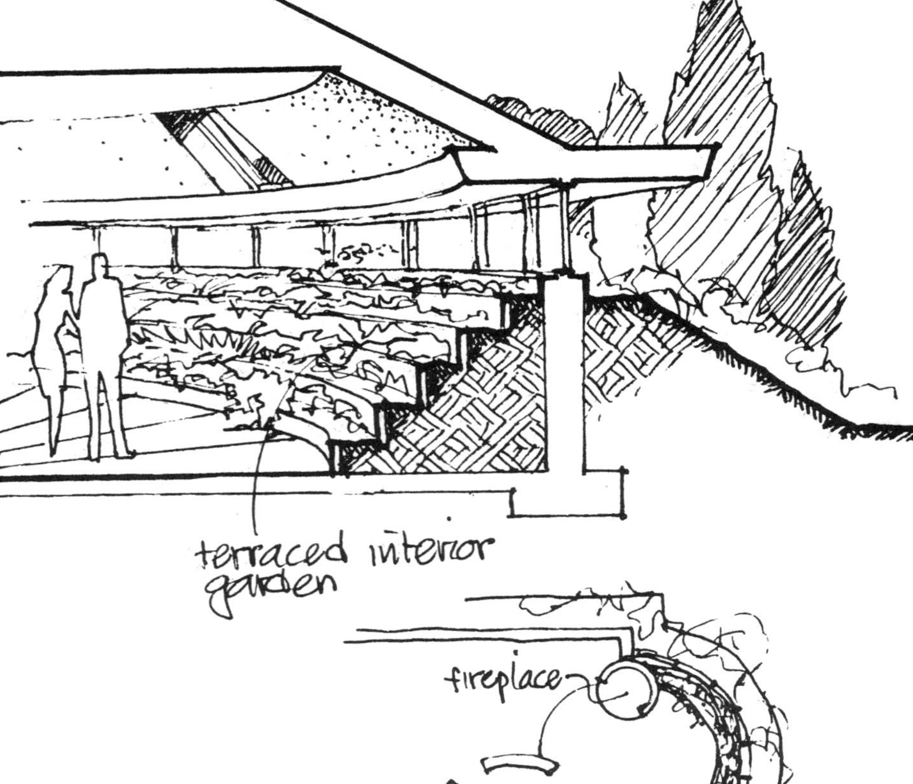

PLAN

Inching toward environmental purity.

Potpourri. Charles almost wraps it up!

("Most important rules in the whole book!" – CGW)

★ important.
★★ very important.
★★★ most important.

1. Do not mix styles. ★★★.
2. Use modules. ★★★
3. Make garage match house. ★★★.
4. All four sides of house count. ★★★
5. Basement low to ground. ★★★ (Probably most important rule in book.)
6. Consistent use of windows. ★★★
7. Proper overhangs. ★★
8. Consistency of design and details. ★★.
9. Use few materials. ★★ (As few different materials as possible, I believe he means.)
10. Proper placement of, and materials for, chimneys. ★★
11. Proper massing and scale. ★★★ (Determined by average human size.)
12. Conceal satellite dishes, TV antennas, etc. ★★★
13. Integrate ells on house. ★★★
14. Proper use of color. ★★★
15. Proper septic mound design. ★★
16. Non-steep front steps. ★★★
17. Properly prepared plans. ★★★
18. Engage a contractor only under a detailed contract that's been checked by your lawyer, and never, ever, make a change without a change order. If you do you will lose $ or end up in court and lose even more $. Honest. ★★★
19. Align & match upper with basem't wdws. ★★★
20. No triangles in windows, doors or wood lattice! ★★★
21. Dog houses etc. should match. ★
22. Site house properly. ★★★

This is Charles's digest of the book. Save it and take the rest of the book to your neighborhood recycling center.

Potpourri. It may end on this very page.

23. Hire an architect. ★★★. It will give you a better house with less (sic) problems, and possibly save you money.
24. And, most important of all: hire me (remember, this is charles speaking). It will do all of rule 23 plus let me pursue Natural Architecture.

As I said, I haven't met Charles yet but I'll bet he wears all natural fibers and even eats raw vegetables on occasion.

Ciraolo Project - Condominiums

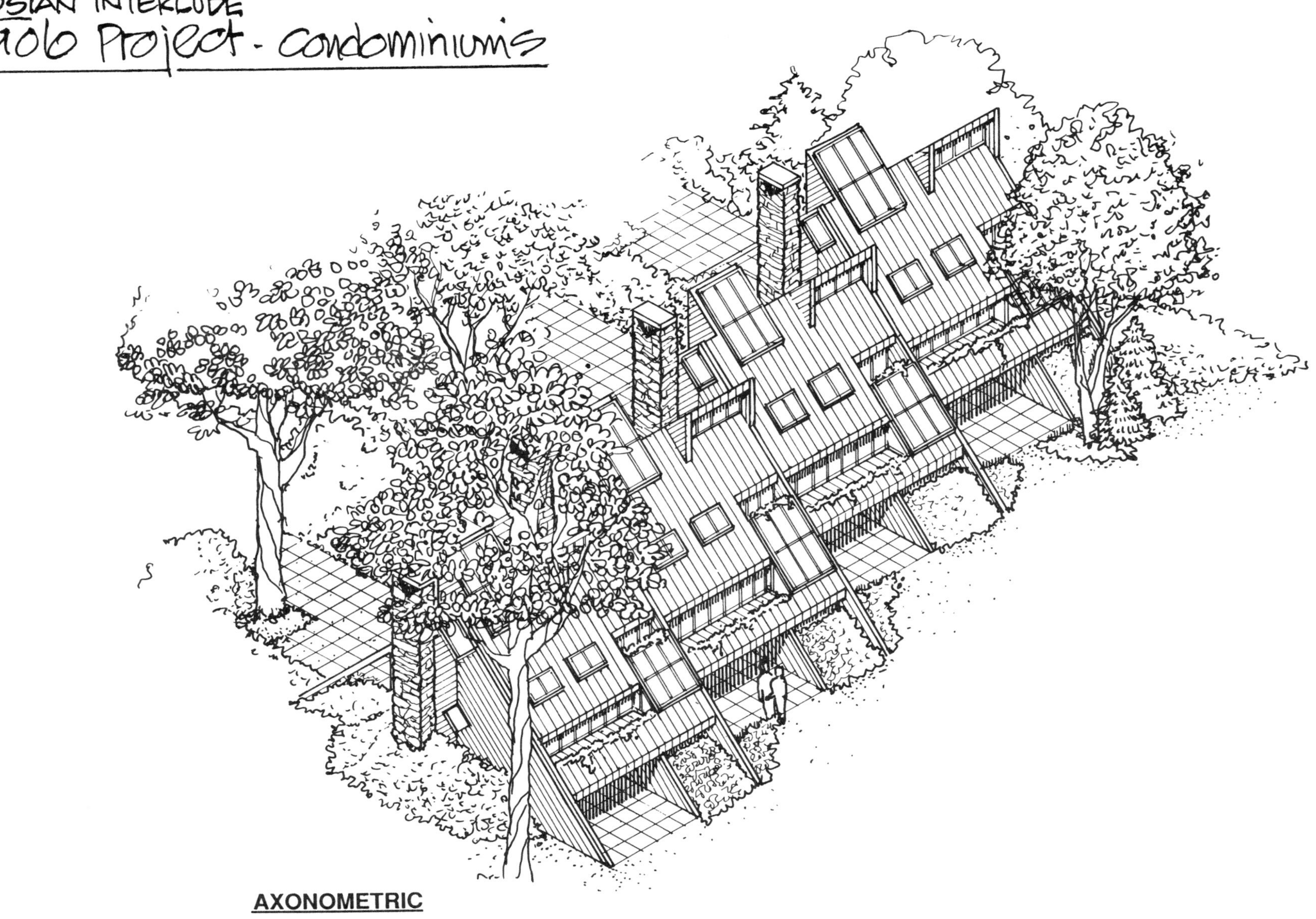

AXONOMETRIC

Rendering by Tracy Boyle.

Designed in 1984.

Why hire an architect?

Charles Woods speaking:

From the Bible, and Aristotle, to the present God has been called an architect – 'The Architect.' An architect does not create "out of nothing" but he does "create" – he imposes 'form' on 'matter', or higher form on lower form.

Too metaphysical? So what? My point is that a single mind – God's or the architect's – has an idea, a single but complex idea. Once he has this idea he must get it built properly. If he does, he says 'it's perfect!' That has not happened to me yet but that's not my point. What is? If someone did a painting or sculpture for you you wouldn't touch it, would you? Or if you were having surgery you wouldn't ask to help, would you? ... or modify your engineer's calculations on that steel beam? And yet people often want to add to, or perfect, their architect's work. An architect is an artist and a scientist. So is it wrong to tamper with his design? No. It's understandable. Our house is like our second body, and, again in the Bible and other religious and mystical writings, the word house

Charles Woods UNFAIR to us atheists

"Not so, Mac! Some of my best friends are atheists... and I've studied the Father of Modern Atheism. Really he's a 'pantheist'!" You brought up the subject." – CCW.

is often used for the soul's body.

We love our houses and want them to represent us. I understand that but, if I am honest, not all changes and improvements made without the architect's approval will help the design. Why? Because a house is a complex idea. Clients – owners – can have great ideas but they must go through the architect's mind if they are to stay within the spirit of his design.

An example: to tell the architect you want a big window is fine. It helps him design what you want. But to say it must be *this* bay window greatly limits the entire concept, not just this window. It's much better simply to suggest that you'd like more light – or view.

Many people say they designed their houses and then "had an architect to draw it up". In some cases the owners may have been better designers than their architects. Most of the time, however, that isn't

Don't forget to tell your architect that there is a great worldwide environmental crisis. He may not have heard about it. I mean, if the President...

true. I don't say this to offend homebuilders but to help them.

If you want a draftsman, hire one, not an architect. But if you want a good design hire an architect. Look at her work, talk to her past clients, and then trust her as you would a doctor or an engineer. Give her all your ideas and then say 'synthesize them; do your best.' The resulting house will usually be worth more than her fee added to its cost.

When someone compliments you on your new car you don't have to be its designer in order to feel pleased. Feel proud of your architect's design, too.

Another example: Frank Lloyd Wright, the great American architect, did many beautiful houses but it was only a client named Kaufmann who said, "Do what you want. We need four (or whatever) bedrooms. We need a summer house. Here's a lot we think is nice. If not, pick your own." The result, "Fallingwater," is considered by many to be the most beautiful house in the world, and it speaks well for Mr.

...and the governor and most other politicians haven't heard about it why should we expect a mere architect to have heard?

Kaufmann as well.

To be honest, if you want a great house hire a great architect. If you want a typical house, design it yourself... or hire a builder to do it. If you want a good house read this book and then design it yourself.

If all this has sounded unnecessarily harsh I really am sorry. But we are drowning in a sea of ugly architecture.

Good job, Charles. I hope they listen to you.

Mystery House

Charles explains: "Here is a case in which people chose the worse over the better. The top version is my cleaned-up view of a design presented to me by some clients. It was one of my rare act-as-a-mere draftsman jobs.* Then I showed the 'potential' clients how I would re-do it (below).

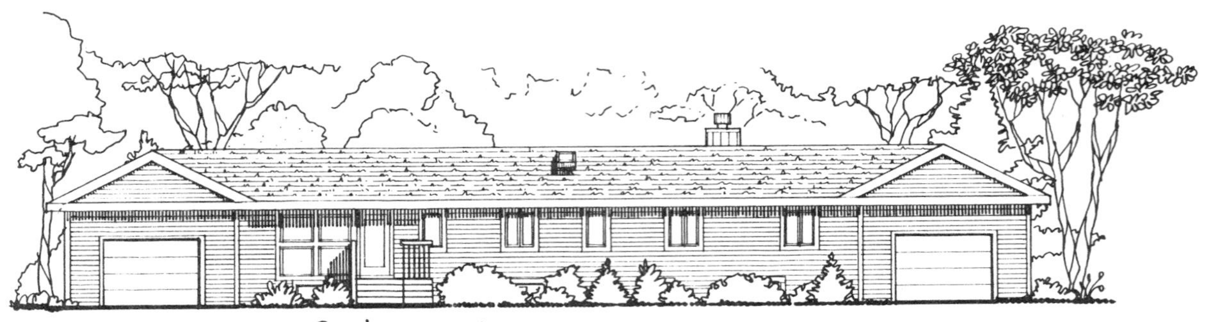

FRONT ELEVATION Drafted version.

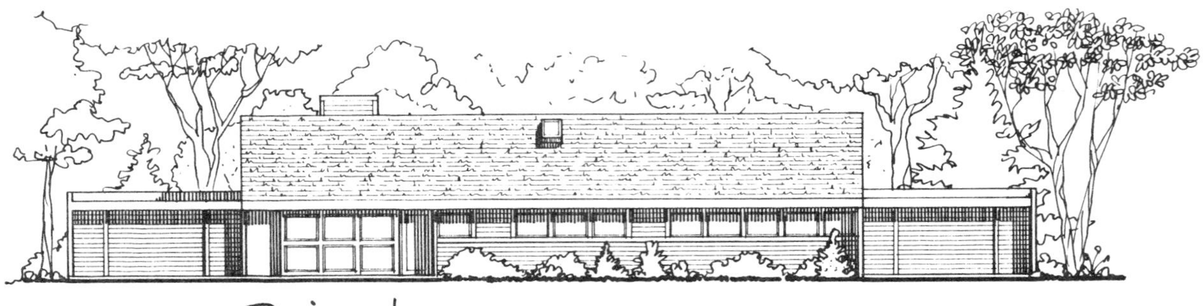

FRONT ELEVATION Designed.

LEFT ELEVATION

The wife liked the re-done design; the husband liked the other one (top). I don't get it!"

*"I was broke!" —CGW

If you can reduce the amount of construction material used in your house by using what others might regard as scraps you get an extra gold ☆ star.

If you don't hire an architect (and Charles and I never did)

Use good stock plans.*

Your builder will be quite happy to dust off the plans of a house he's built with great success many times in the past but chances are you'll find more of what you want in drawings created by professional stock-plan designers. Then, using our jim-dandy book as your guide, you can modify even that worthy design until it fits you well. Then try to enjoy the frustrating, disappointment-ridden, wild, wonderful experience called building your house. If your sense of humor is intact you'll have a great time. Unforgettable.

In this energy-waster the living room is a passageway and the foyer and stairs are cramped...

...while here the house is energy efficient as well as part of its living site. And the foyer/stair hall is at least a bit more gracious.

*see rear of book for some of ours. Charles** says he will mark up anyone's "plans, etc." for $100. (mark up, not draw up!) I was tempted to undercut him by $5.00 but his is a real bargain.

**C.G.W. / R3, Bx 538 / Honesdale, PA 18431

America still has several million acres of farm and forest land, so let's get out there and cover them with new houses.

"Do not build off the back of a napkin," says Charles.

Good advice.
Build off the front instead, and you'll stay out of trouble.

No? Got it wrong again, have I? Well, let's analyze his statement and figure out what it means. He's probably trying to say that a carelessly drawn plan, while it may contain the germ of a good idea, needs to be worked out in some detail in order to be sure that the design is consistent throughout the house and that all the parts will fit together. Without plans & specs you won't know what you'll get, and you'll have no legal protection.

From the paper napkin you got this

when you thought you were getting this.

If a house seems to harmonize with its site chances are that it really does. If you plan it that way, all the better. Good luck!

Consider hiring a professional to look over your plans.

Your local building inspector will be helpful in checking to see that everything meets the requirements of the building code but it is the rare inspector who will check and approve your structural design. It's not his job to certify your structure. It's your responsibility, and if you are not well-versed in using the structural formulas and tables associated with building structures, spend the money to find someone who is.

A local architect or structural engineer will probably look over your drawings and make helpful comments in return for a modest fee. But if you want her to certify the design by putting a seal on the drawings you must expect to pay a lot more for such security.

Remember what happened to Clarence Poddington, the man who coined the phrase, "I don't need no pro to tell me how to build."

When you talk to an architect or an engineer make it clear that your environmental priorities are high. Don't let them get trampled.

The Construction Contract

The rule: make sure there are no loose ends before you lift that pen. "And then make sure again," adds Charles. You should have detailed plans and specifications. Be sure everything has been spelled out. Get a lawyer involved. You're about to spend a lot of money, and when things go wrong you can be pinched for years to come unless you're protected. You also want to protect the builder from a failure, on your part, to pay for authorized work, so let fairness be the rule.

One of the best ways to be sure that a builder will do what's required of him, and do it in the allotted time, without hassles and claims for "extras," is to pick someone with a good reputation. How? By interviewing at least two of his recent customers. Listen carefully to what they say, then decide if he's someone you want to bid on the work. You should get from 2 to 4 bids and then award the contract to the lowest bidder. Don't play games with them after they've done all the work of bidding.

The surest way to increase the cost of a house is to make changes during construction. It also gives the builder a way to manipulate you. Restrain

Did you include a composting toilet in the design? ... leave room on the sunny side for a vegetable garden ... leave some area wild?

The construction contract (cont.)

yourself! You can make some of those changes after you move in. The best practice, of course, is to study the plans so well you're able to make all the changes on paper before the contract is signed.

After reading, and revising, the foregoing, Charles added the following thoughts:

Important! Also - I suggest using an AIA construction contract (available at any large stationer's), and I urge you to remember that a 4-to-6-month completion schedule will almost always be 8 to 12 months, minimum.

Have someone - an engineer? - supervise the construction. Check it yourself, too. Read a few books on the subject, get involved.

Do I sound like someone who doesn't trust builders? I love a good builder. They are very rare.

Make sure, by getting receipts, that the builder is paying his bills. Don't overpay him! Why should he work if he has most of the money in advance? I have seen great suffering of both homeowners and builders because of ignoring these rules.

Be tough-minded in seeing that areas of the site marked "not to be disturbed" are actually not disturbed. Strings and ribbons won't do it. Use sturdy fences.

South Elevation

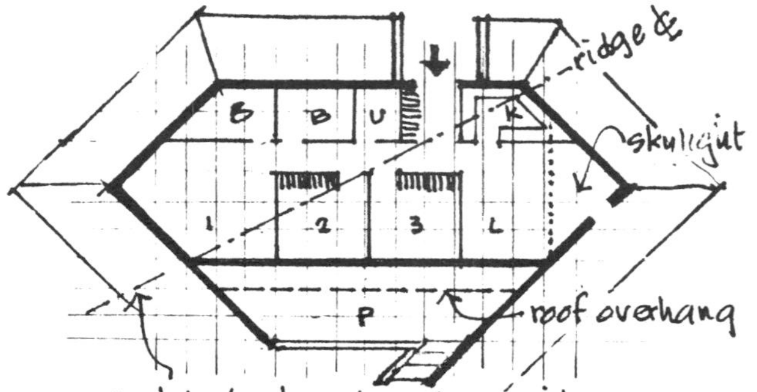

And look where he put the ridge of the roof! It's not even on the 45° angle of all the end walls but it <u>works</u>. He knows what he's doing, that young man.

Charles is preparing his own designs for presentation in this book but I thought I'd surprise him by doing this one from some big prints he gave me. I like his control of the design. It's pure. But after several tries at an aerial perspective I gave up. I just couldn't make that folded roof plane look convincing. I'm too used to doing rectangles, I guess.

Plans available with detached matching garage.

I think that a condition of geometric receptiveness in the brains of designers leads them to grow crystals like this. —M.W.

The Small House
Rendering by Mac
Heavy-handed Mac.

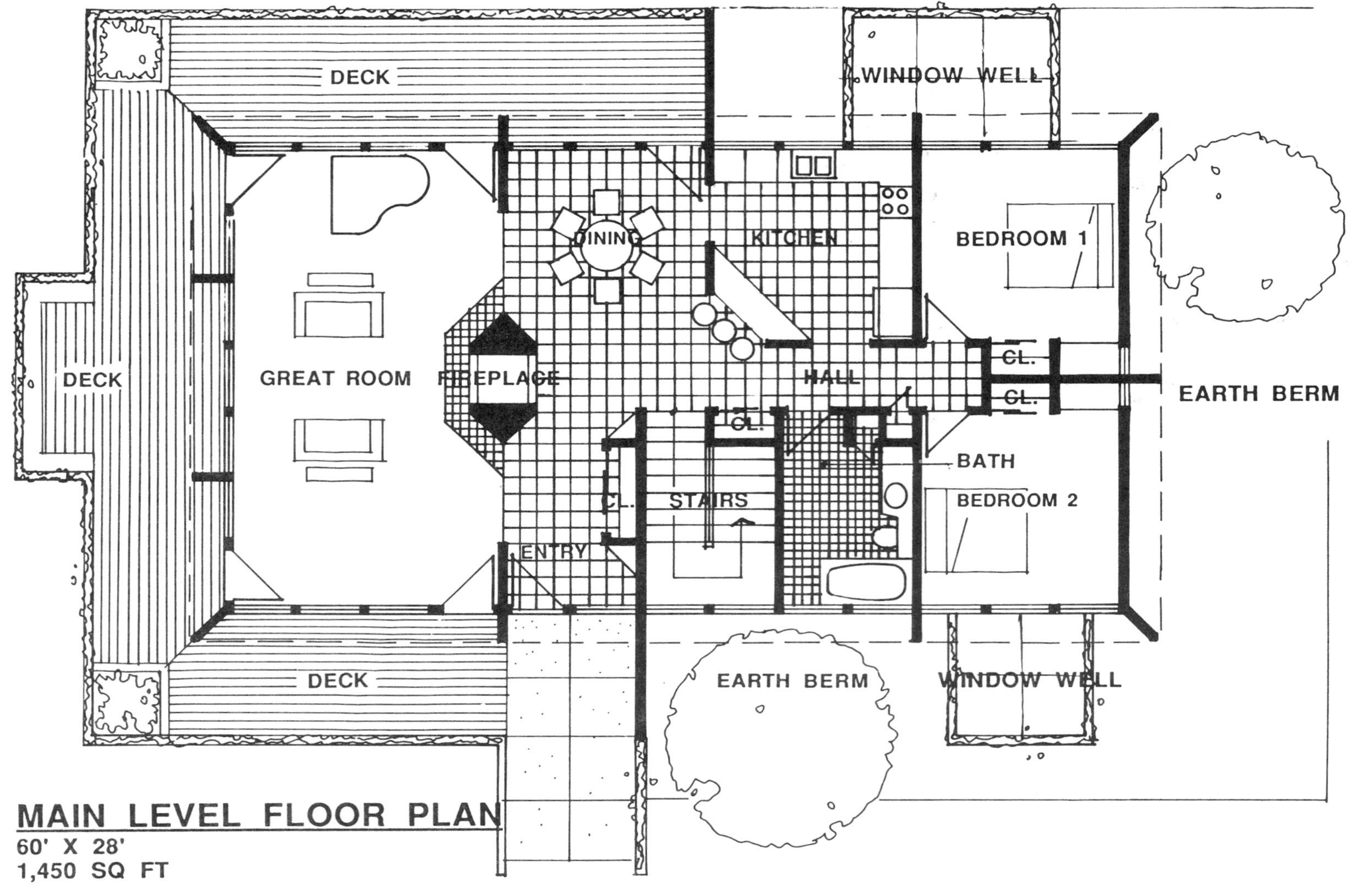

Small Residence (basement plan not shown.)

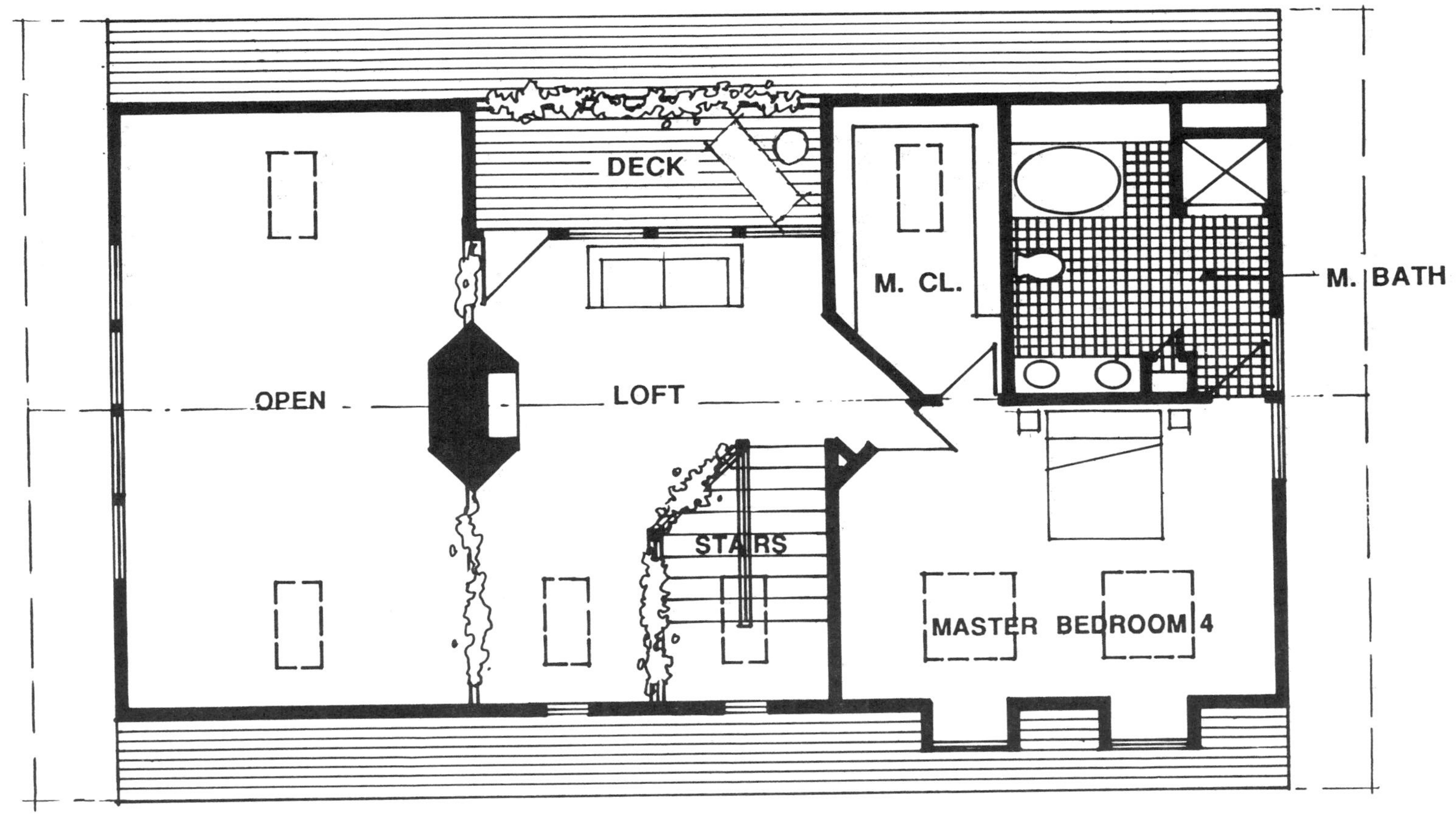

UPPER LEVEL PLAN

1,050 SQ FT

Small Residence.

The Augello House, Honesdale, Pa.
"My town," adds Charles.

Built by longsuffering general contractor, Jerry Dulay -- for $140,000.

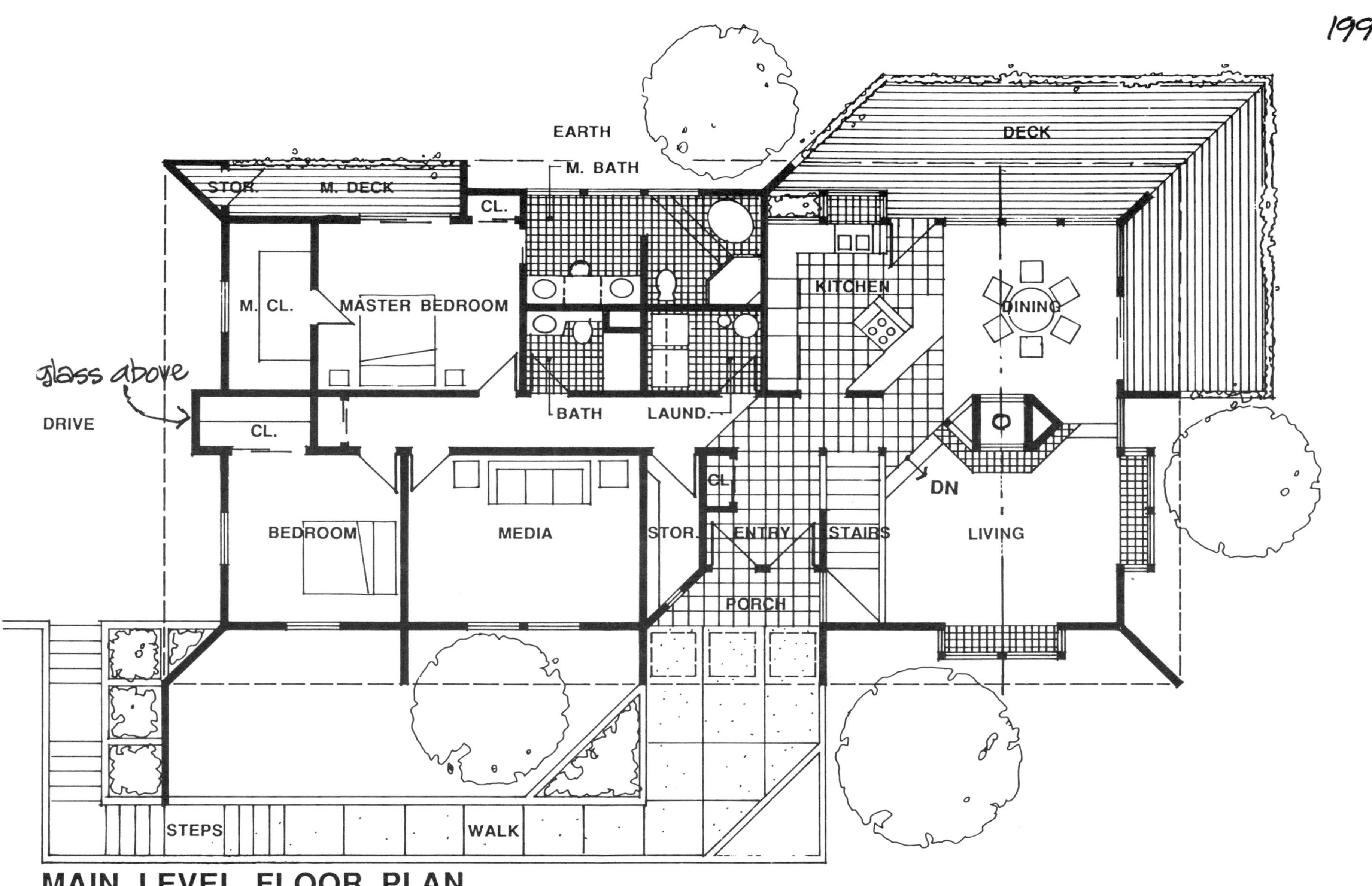

MAIN LEVEL FLOOR PLAN

52' X 28'
1450 SQ FT (per floor)

Augello Residence (basement plan not shown)

South perspective

Rendering by Randy Black

"Omega"

Construction plans, as seen in POPULAR SCIENCE, are available.

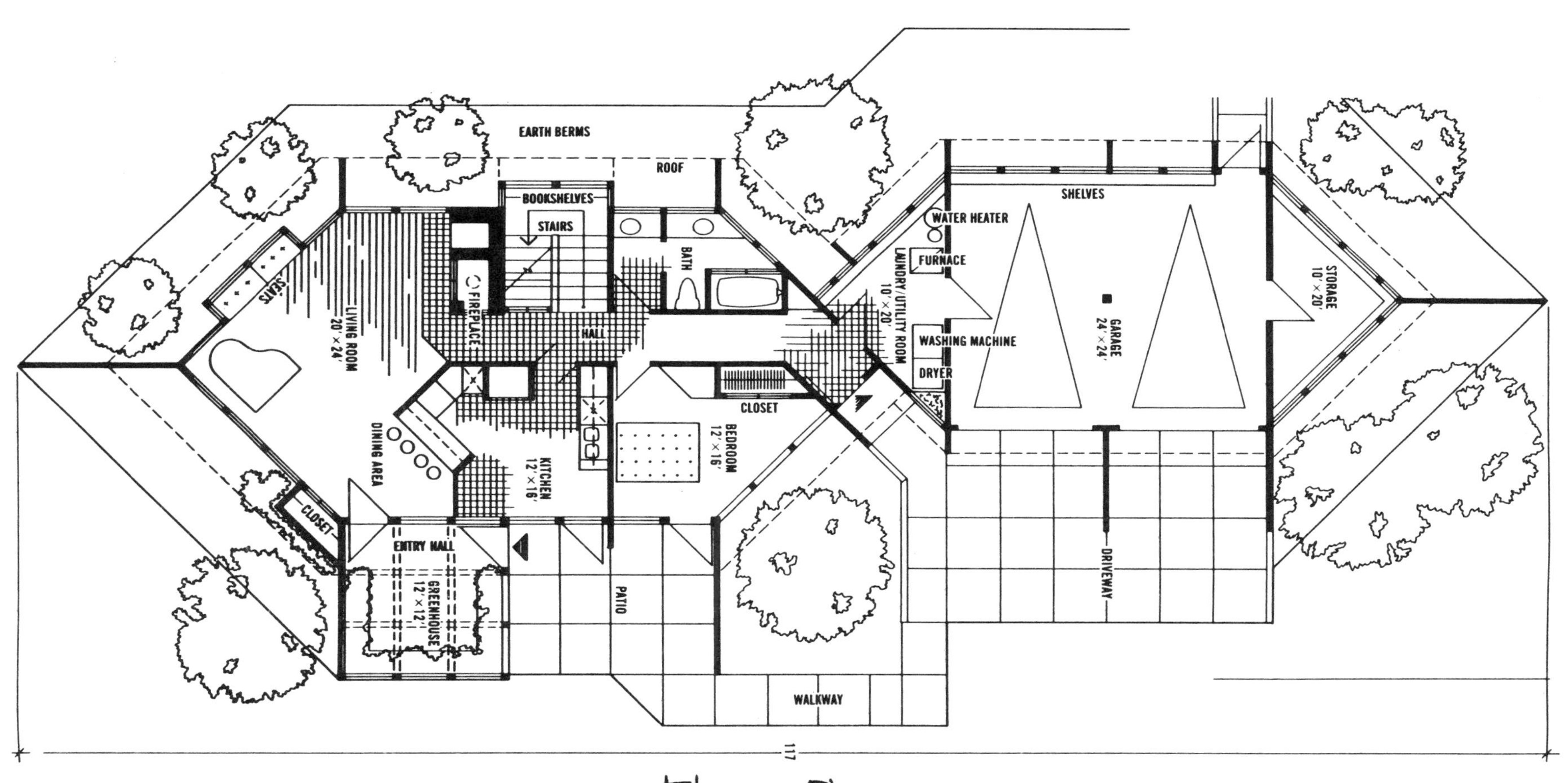
EARTH BERMS
ROOF
BOOKSHELVES
STAIRS
BATH
SEATS
LIVING ROOM 20' × 24'
FIREPLACE
HALL
WATER HEATER
FURNACE
LAUNDRY/UTILITY ROOM 10' × 20'
WASHING MACHINE
DRYER
SHELVES
GARAGE 24' × 24'
STORAGE 10' × 20'
CLOSET
BEDROOM 12' × 16'
KITCHEN 12' × 16'
DINING AREA
CLOSET
ENTRY HALL
GREENHOUSE 12' × 12'
PATIO
DRIVEWAY
WALKWAY
117

Floor Plan.
2500 sq. ft.

Omega.

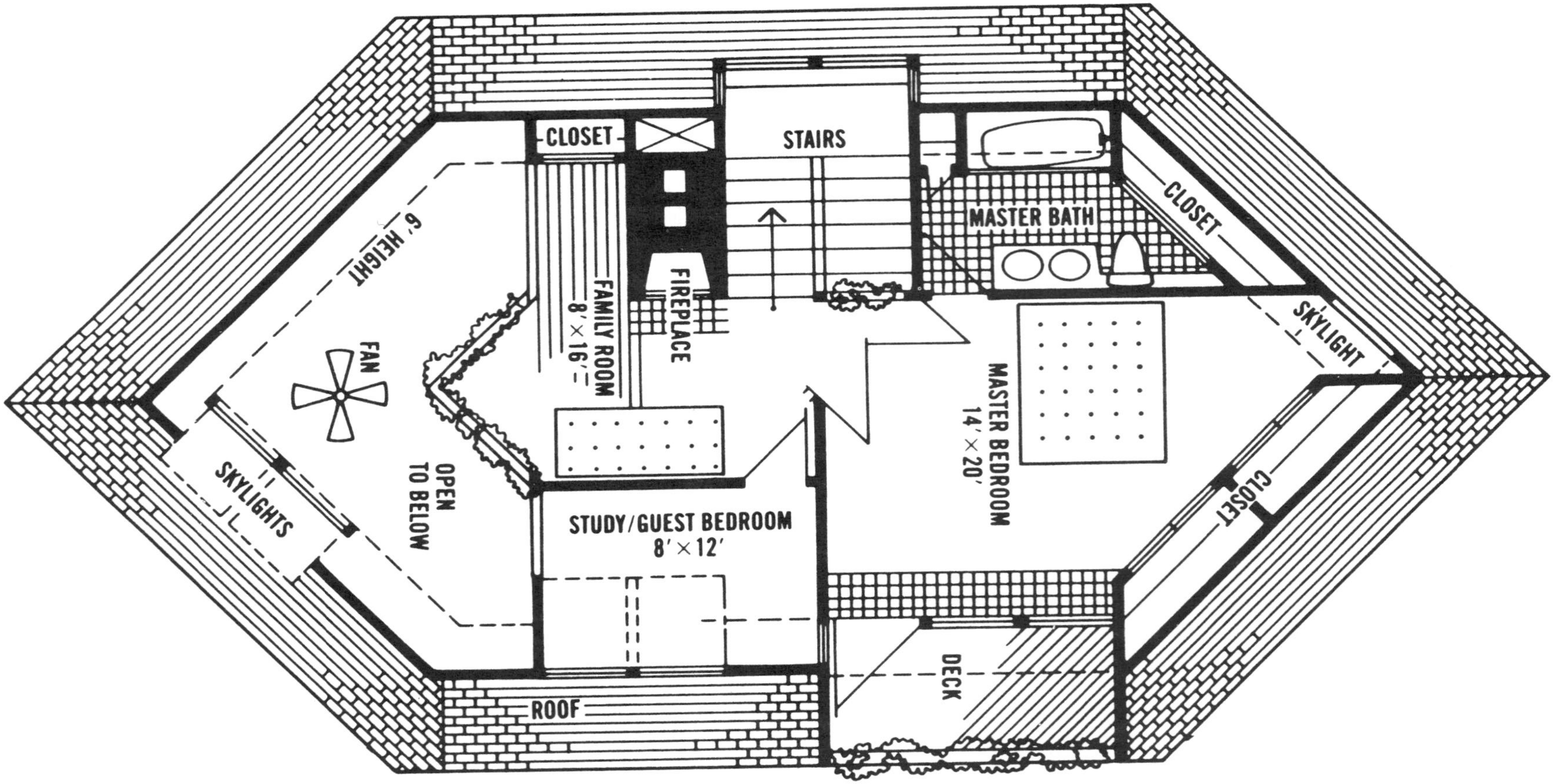

CLOSET
STAIRS
MASTER BATH
CLOSET
6' HEIGHT
FAMILY ROOM
8' × 16'
FIREPLACE
FAN
SKYLIGHT
MASTER BEDROOM
14' × 20'
OPEN
TO BELOW
SKYLIGHTS
STUDY/GUEST BEDROOM
8' × 12'
CLOSET
DECK
ROOF

Second-Floor Plan.

(It reminds the clients of a "Klingon - Bird of Prey" warship from "Star Trek"! Is that good or bad?)

Rendering by Randy Black

Stultz Residence. (One of my favorite clients!) Clinton Twp., N.J.
Construction plans are available.

Floor Plan (4200 square feet)

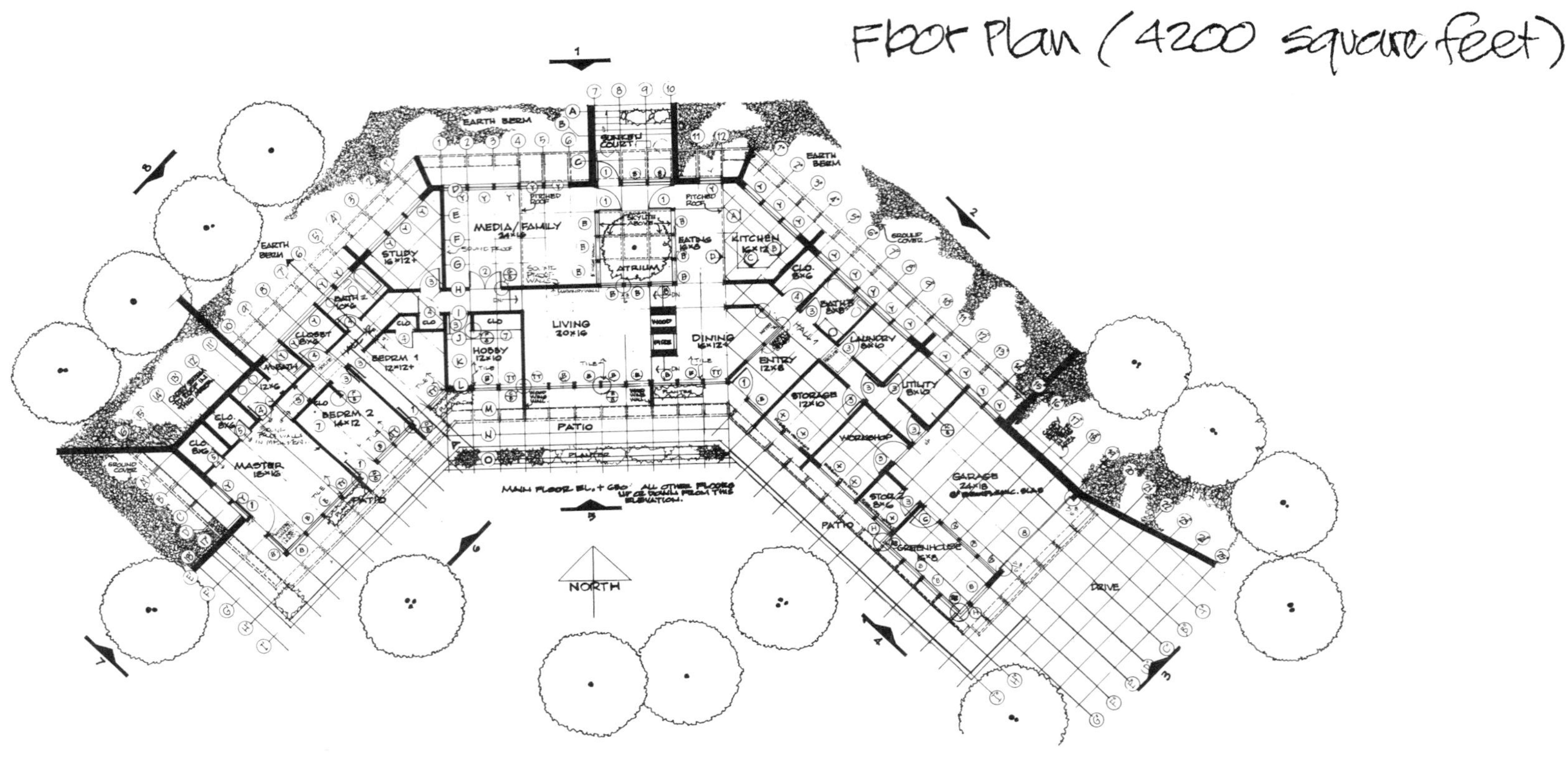

Stultz Residence - currently under construction.

sectional perspective.

Stultz Residence.

south perspective.

A LAKEVIEW RESIDENCE FOR GEORGE & SHARON BYERS
DESIGNED BY CHARLES G. WOODS NATURAL ARCHITECTURE

A great rendering by Jay and Tracy Boyle, Architects.

Byers Residence
Construction Plans are available.

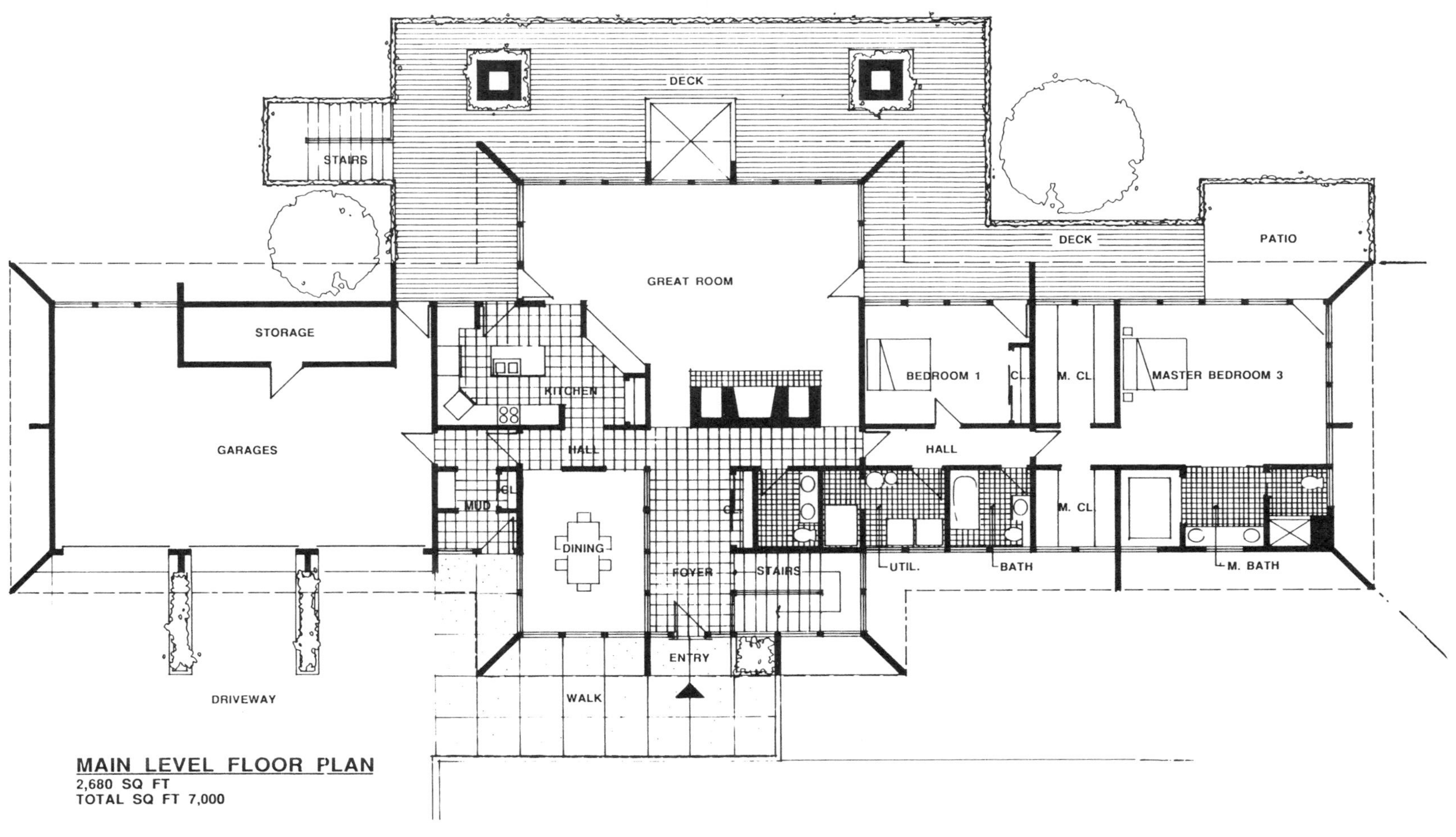

Byars Residence (Basement not shown.)

Although this is an accurate perspective - it looks sort of like it might 'get up and walk'. From street level it would look quite content to stay still and beautiful.

Rendering by the Boyles

Court House - as seen in Popular Science. Construction plans are available.

Rendering by Tracy Boyle

"The Prow."
Construction plans are available.

MAIN LEVEL FLOOR PLAN

32' X 32'
1,024 SQ FT

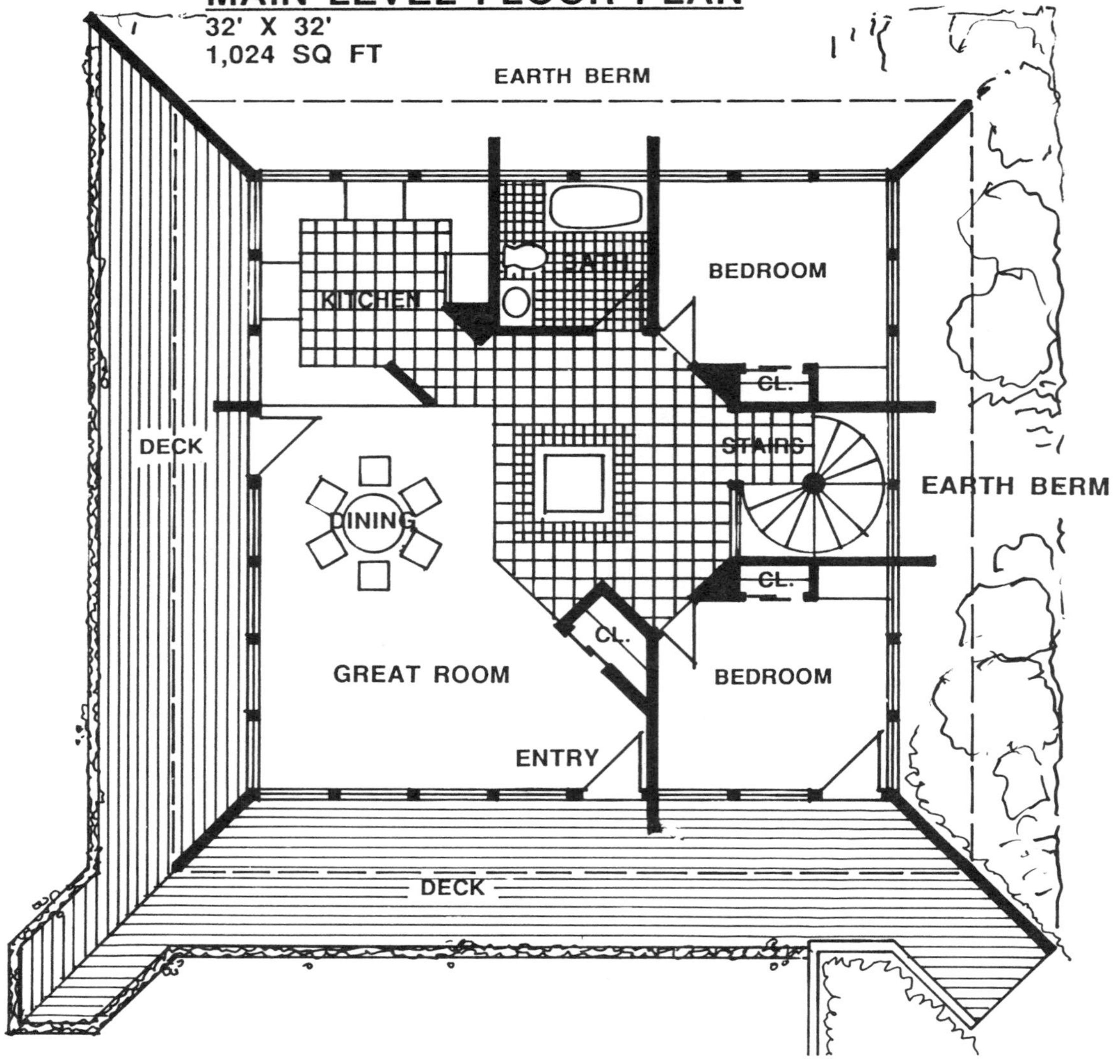

EARTH BERM
KITCHEN
BATH
BEDROOM
CL.
DECK
STAIRS
EARTH BERM
DINING
CL.
CL.
GREAT ROOM
BEDROOM
ENTRY
DECK

STOR.
HALL
MASTER BEDROOM
CL.
M. BATH
CL.
down
STAIRS
LOFT
(open)
STOR.

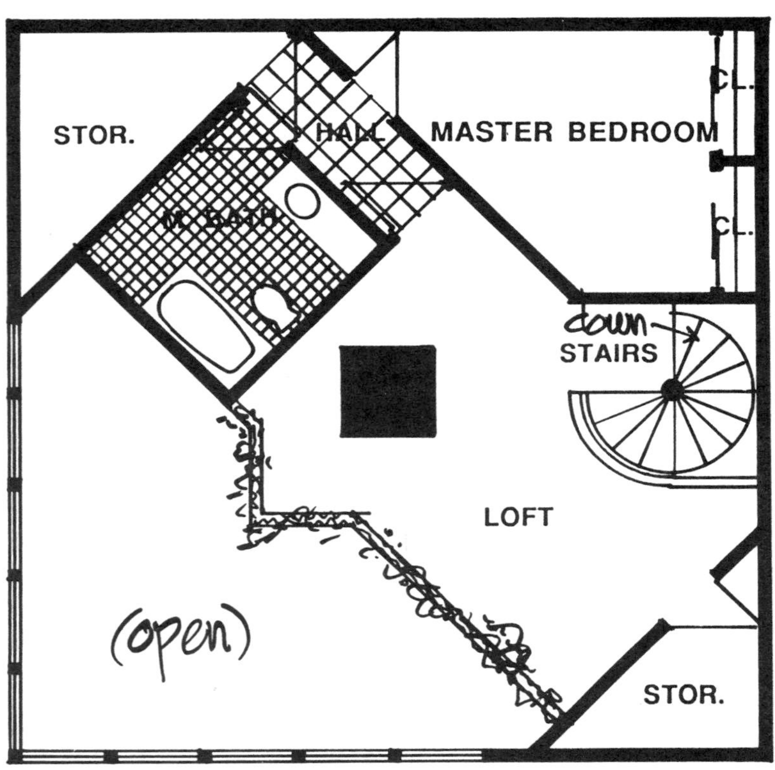

LOFT PLAN

700 SQ FT

"The Prow" (Basement plan not shown)

Rendering by Tracy Boyle

"Sparrowhawk", one of my least expensive houses! It could be built for $80,000. Construction plans are available.

212

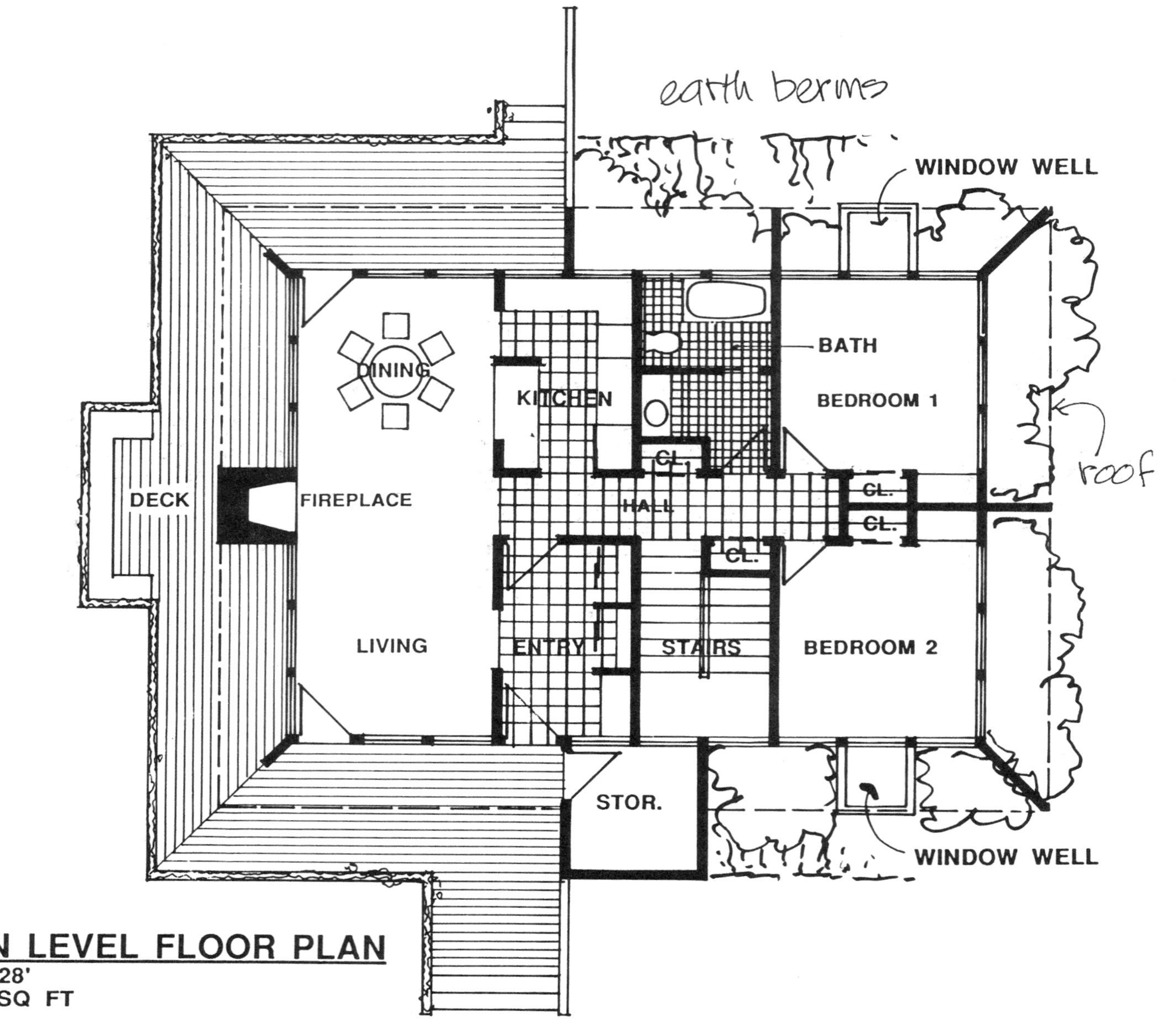

MAIN LEVEL FLOOR PLAN

40' X 28'
1,120 SQ FT

"Sparrowhawk".
(Basement plan not shown: TV/bedroom below.)

Rendering by the Boyles

Radiant House. (One of my favorites! – C.G.W.)
Construction plans, as seen in POPULAR SCIENCE, are available.

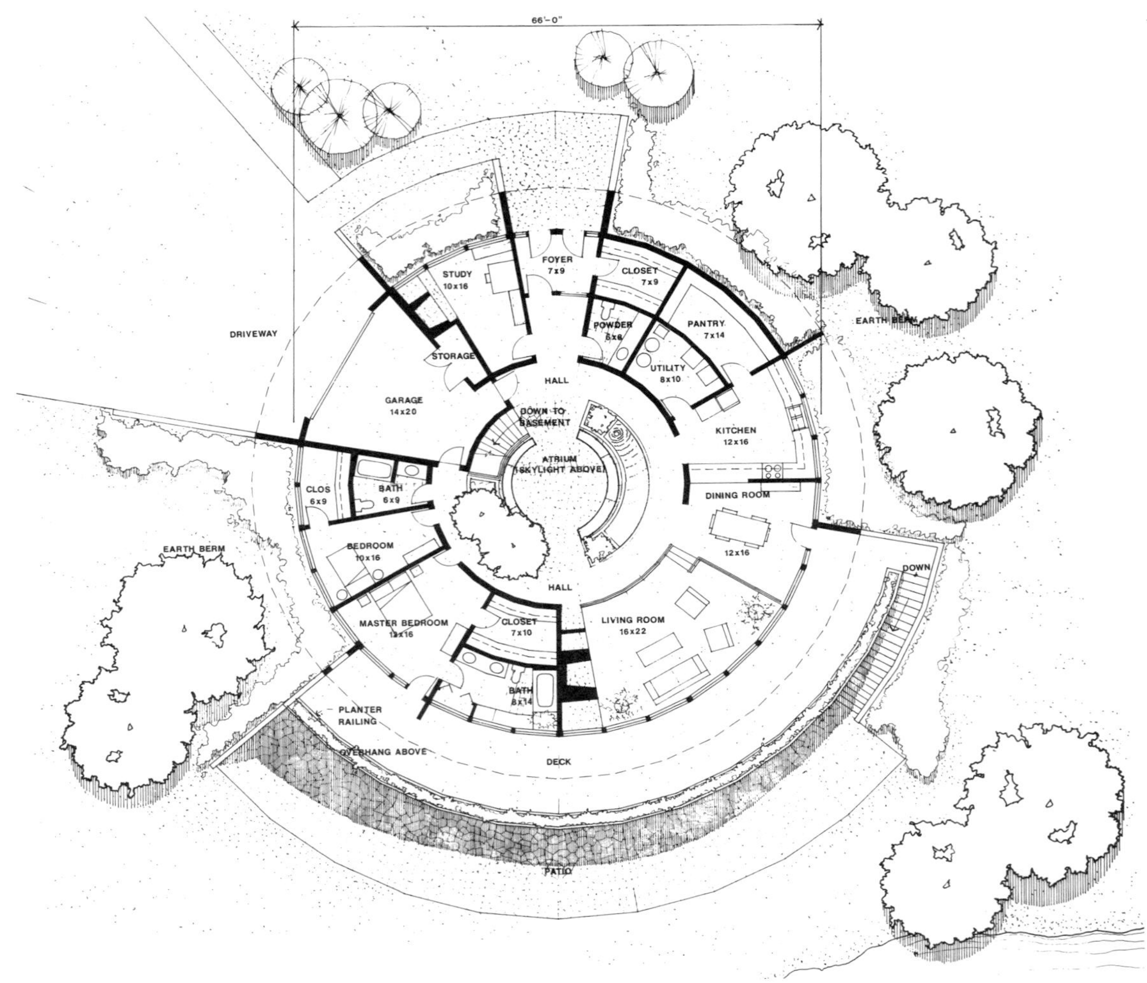

Floor plan. Main floor: 3500 sq. ft.
Lower level: 1750.

Radiant House.

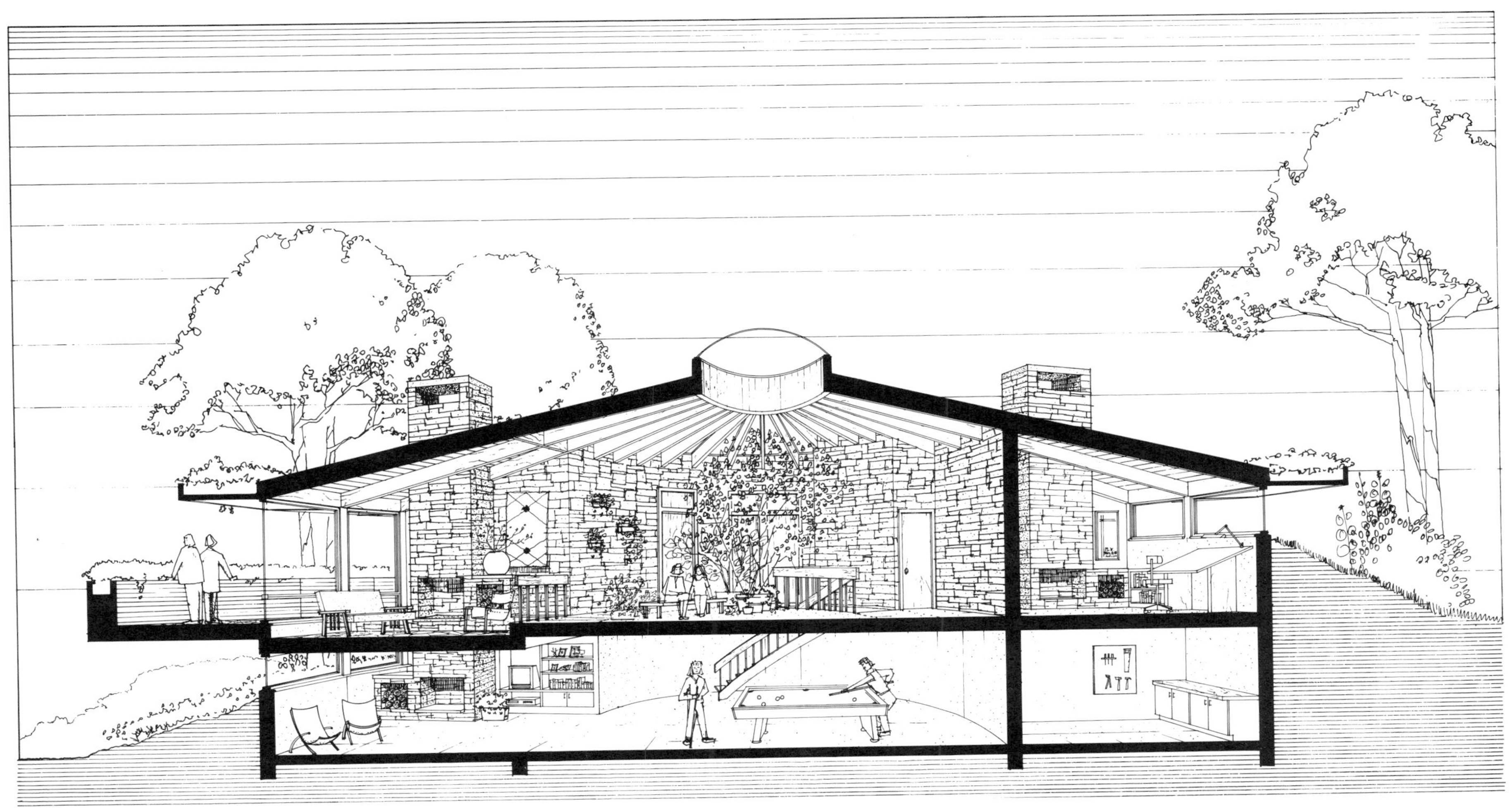

Sectional Perspective.

The Boyles are brilliant!

Designed
By Jay and Tracy Boyles, Architects. (C. G. Woods, consultant)

The Bow House

Construction plans are available through Charles Woods.

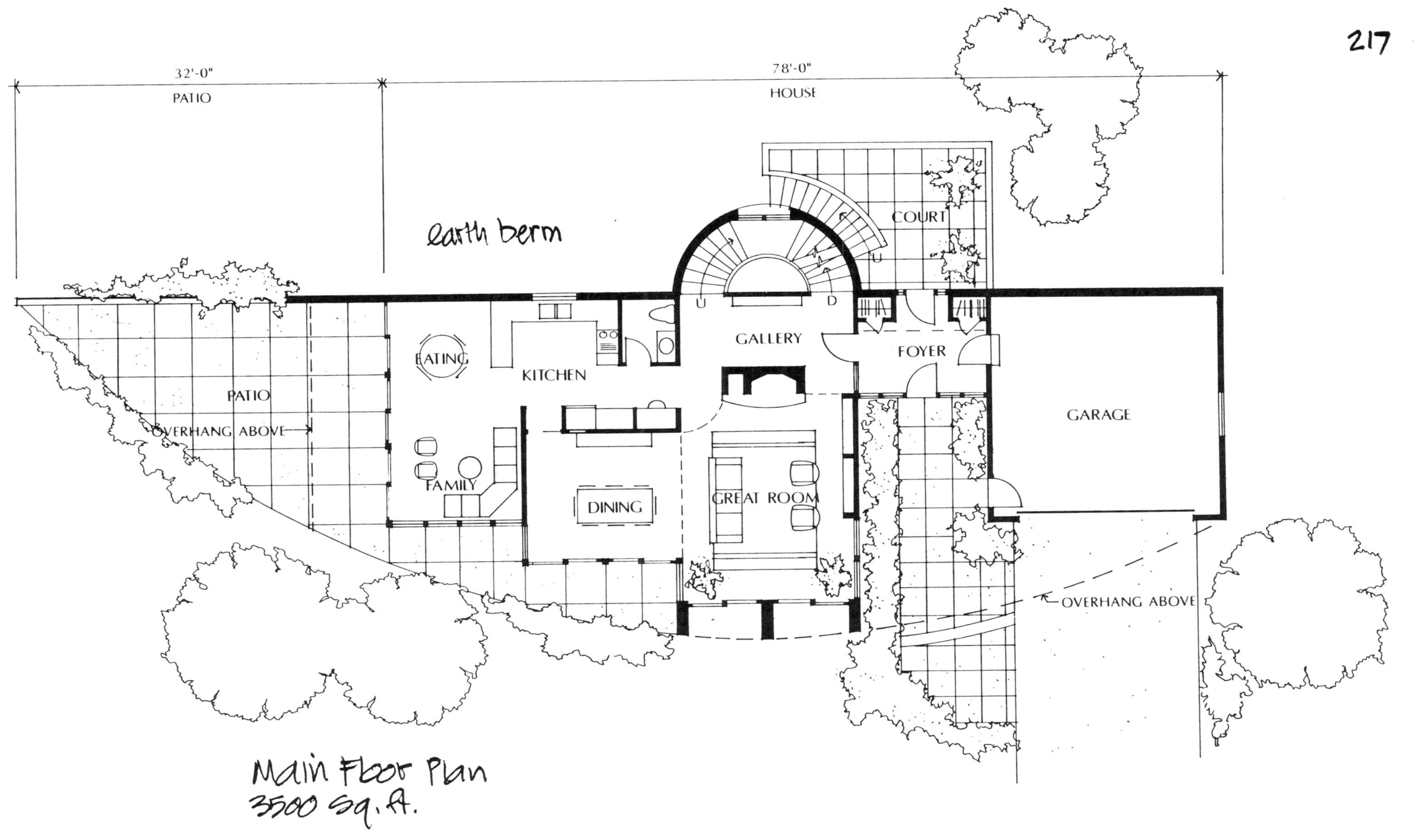

The Bow House . . . an interesting example of mixing curves and rectangles with even a little playful post-modernism.

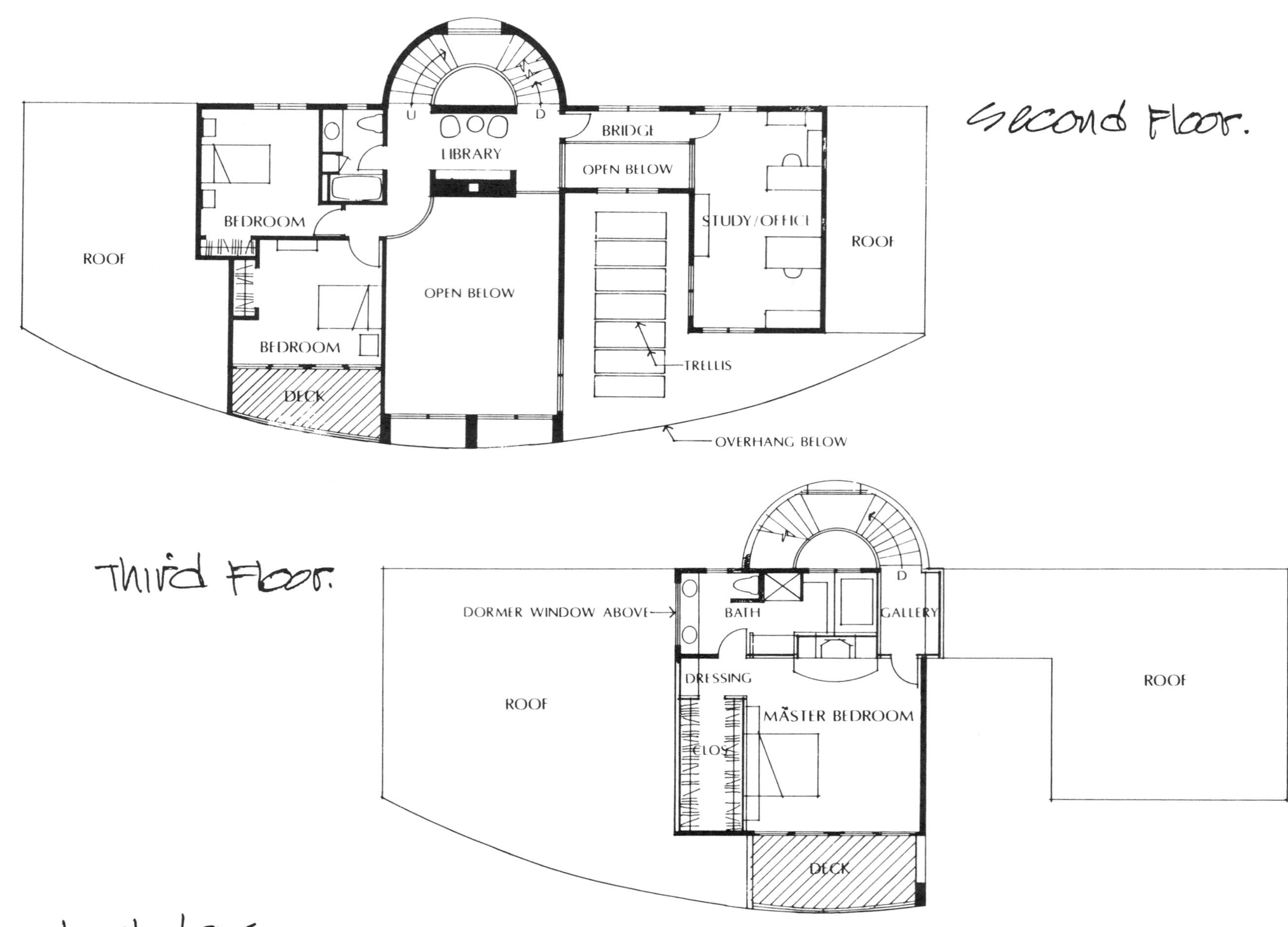

Upper level plans.

The Bow House

The Bow House by the Boyles

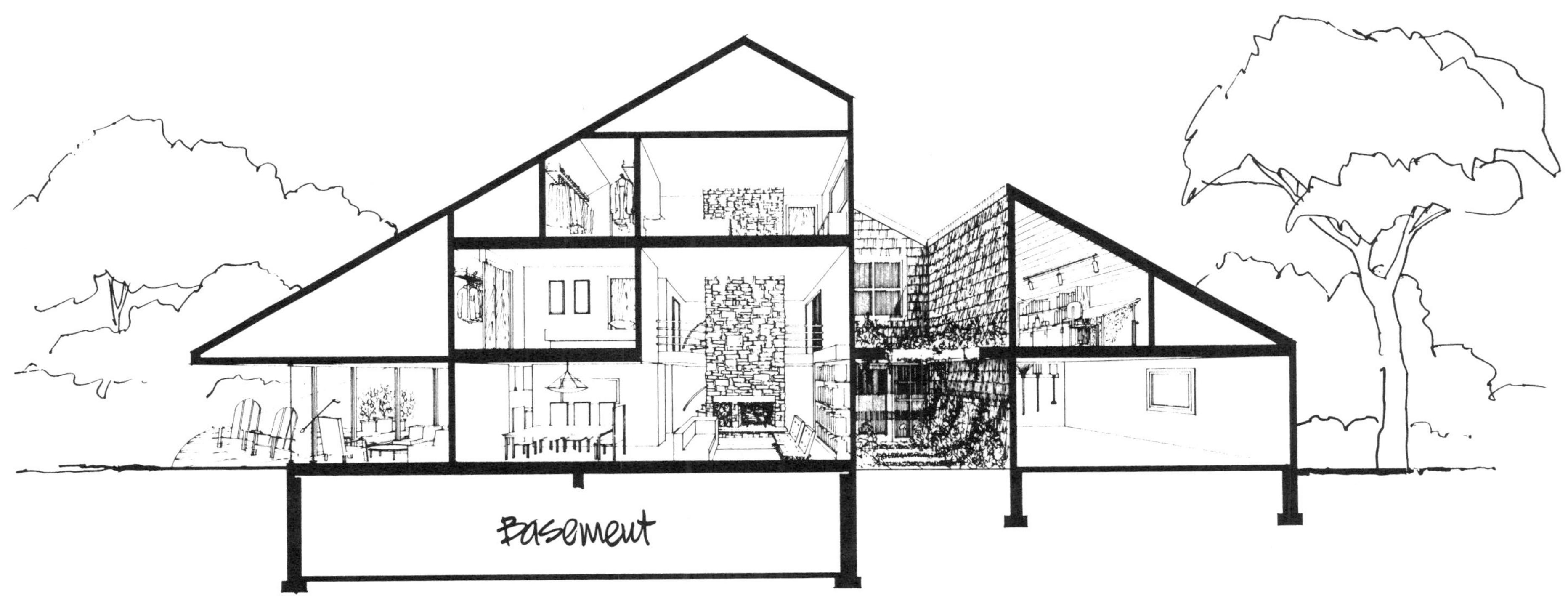

sectional perspective

The Senalik Project — by Charles G. Woods

In this book there has been a lot of hopefully useful information portrayed with humor and cartoons but our goal is not humorous. It is instead serious; safe, beautiful architecture — which is art and science.

We have tried to balance this humor with rendered house designs throughout the book, and complete designs at the rear of the book, but I thought you would like to see one of my construction plans (partial). I chose my Senalik residence. It is one of my largest and most complicated to date. It is also one of my most Wrightian in inspiration and it exemplifies many of the principles in this book.

The house, at 5500 square feet, would cost between five and six hundred thousand dollars but could be built smaller.

What happened to it? The clients decided not to build it in its present form. Back to the drawing board.

Nevertheless, it shows the number of drawings (13 total) and details involved in a well-designed house. (In an actual project, supervision and field details would follow.)

I designed the house and Wright-trained master architect Albert Sincavage was the consulting architect and draftsman.

The Senalik residence is an

The Seralik Project (cont.)

earth sheltered, superinsulated, passive solar house designed in the spirit of Frank Lloyd Wright's 1940s and Usonia houses. It varies somewhat from his houses, of course - most notably in its plank and beam structure and its earth covered roofs.

This is the closest I've come to Frank Lloyd Wright's "Fallingwater."

Rendering by Randy Black

Senalik Residence. The final design is larger than this earlier perspective indicates. The construction plans that follow are available.

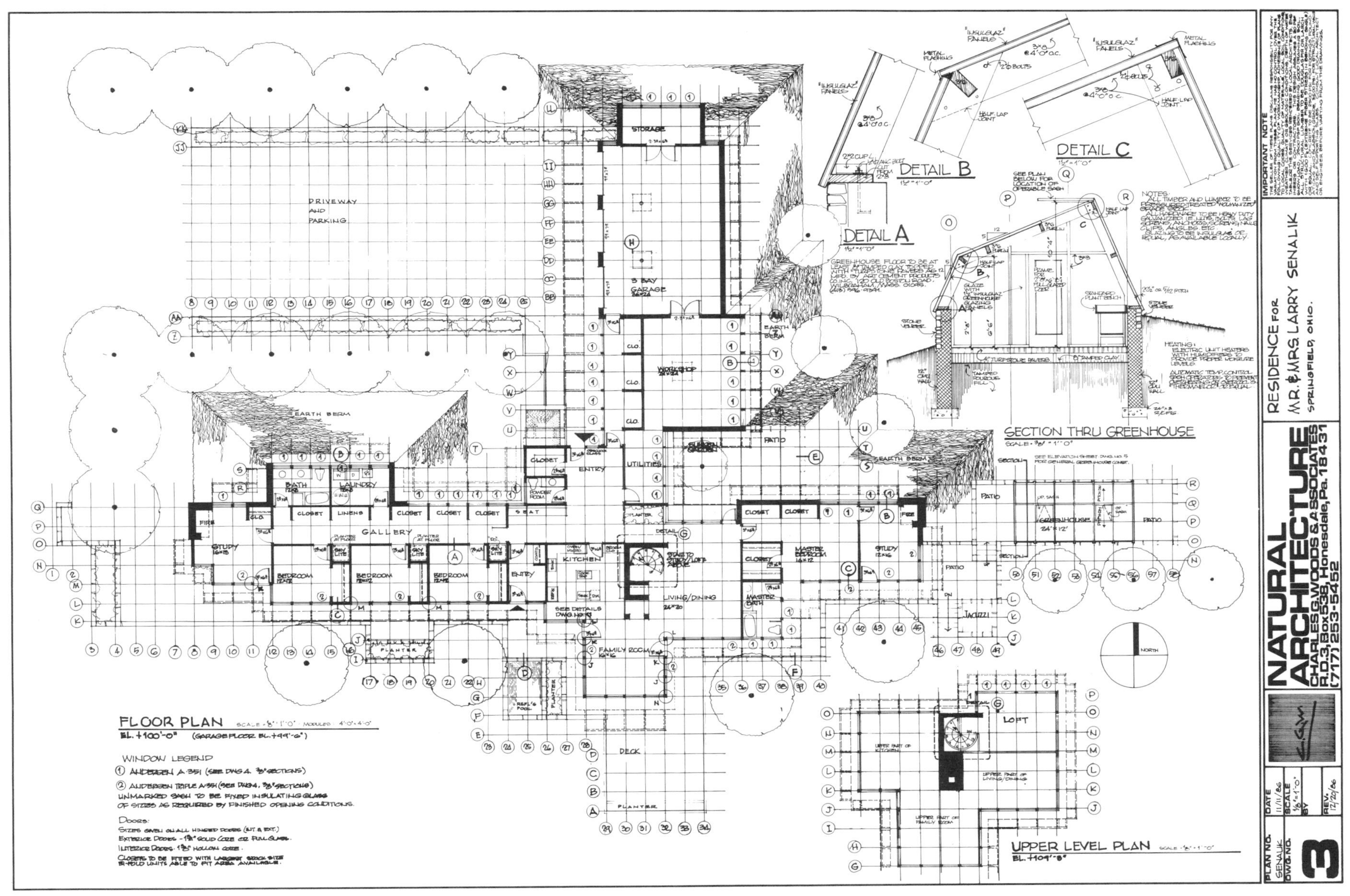

Senalik Residence construction plans (partial).

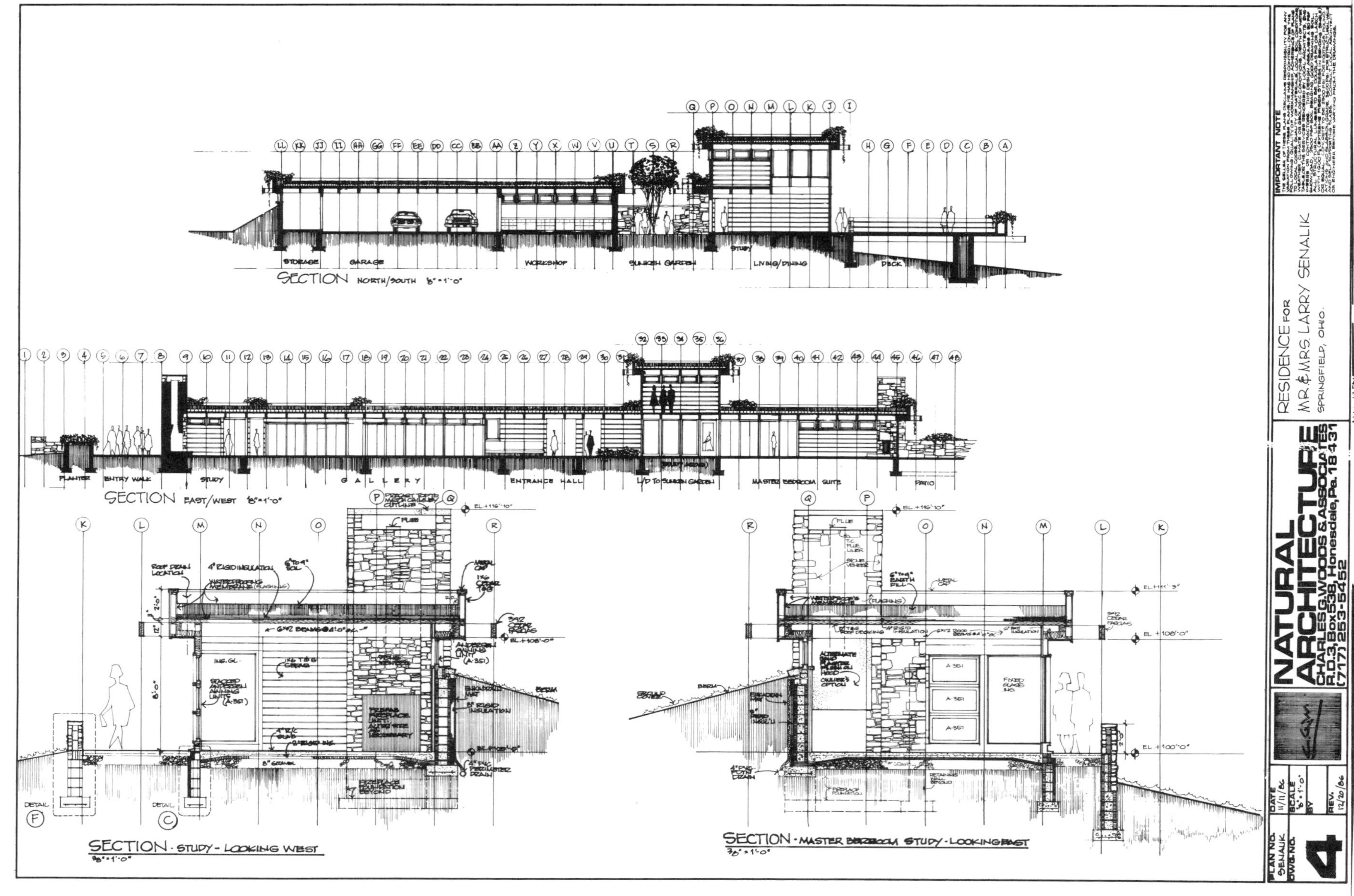
SECTION NORTH/SOUTH 1/8" = 1'-0"
STORAGE
GARAGE
WORKSHOP
SUNKEN GARDEN
STUDY
LIVING/DINING
DECK
SECTION EAST/WEST 1/8" = 1'-0"
PLANTER
ENTRY WALK
STUDY
G A L L E R Y
ENTRANCE HALL
MASTER BEDROOM SUITE
PATIO
SECTION - STUDY - LOOKING WEST
SECTION - MASTER BEDROOM STUDY - LOOKING EAST
IMPORTANT NOTE
RESIDENCE FOR
MR. & MRS. LARRY SENALIK
SPRINGFIELD, OHIO
NATURAL ARCHITECTURE
CHARLES G. WOODS & ASSOCIATES
R.D.3, Box 538, Honesdale, Pa. 18431
(717) 253-5452

Senalik Residence construction plans.

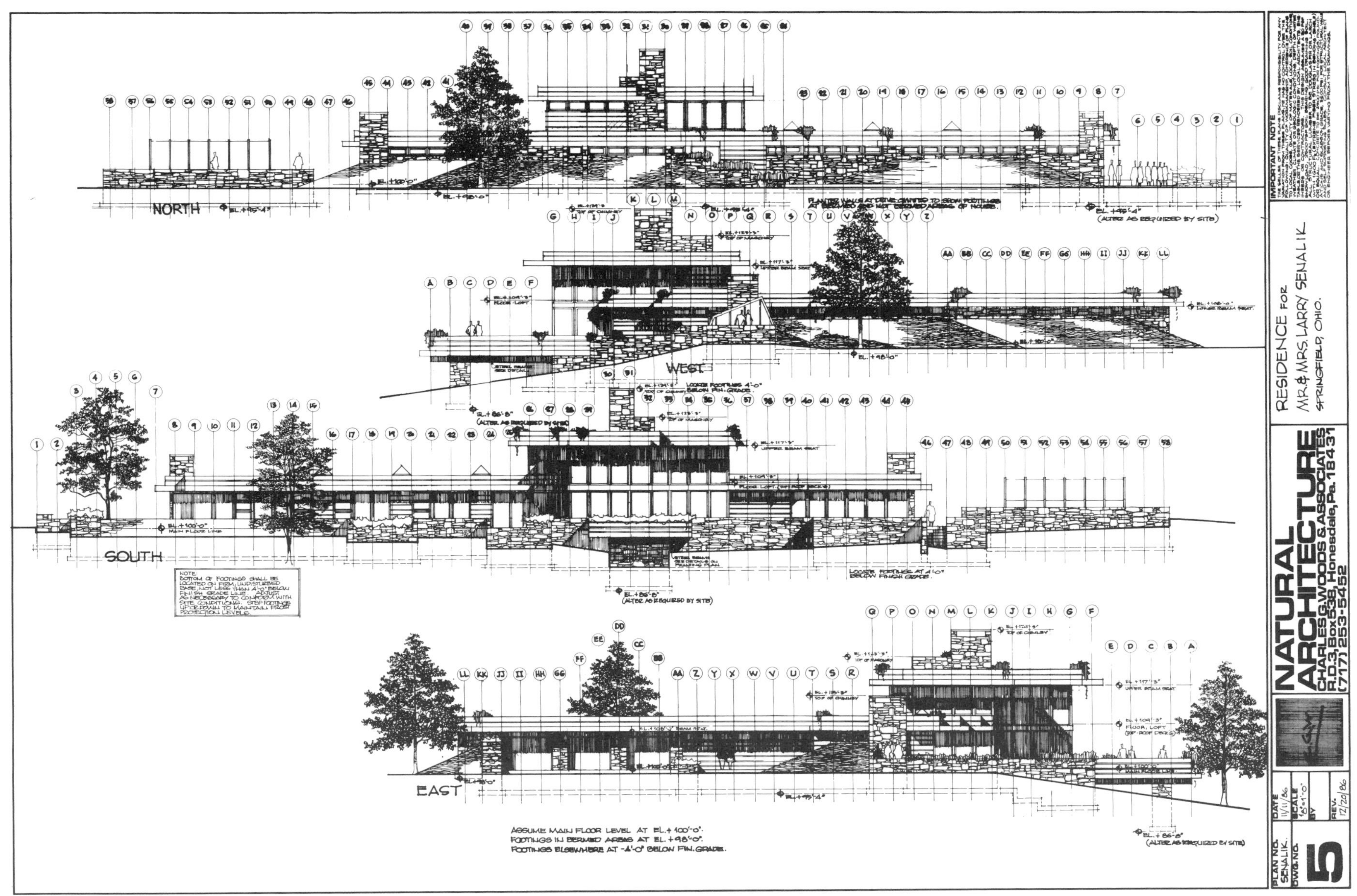

Senalik Residence construction plans.

Please read that upper-right corner paragraph carefully!

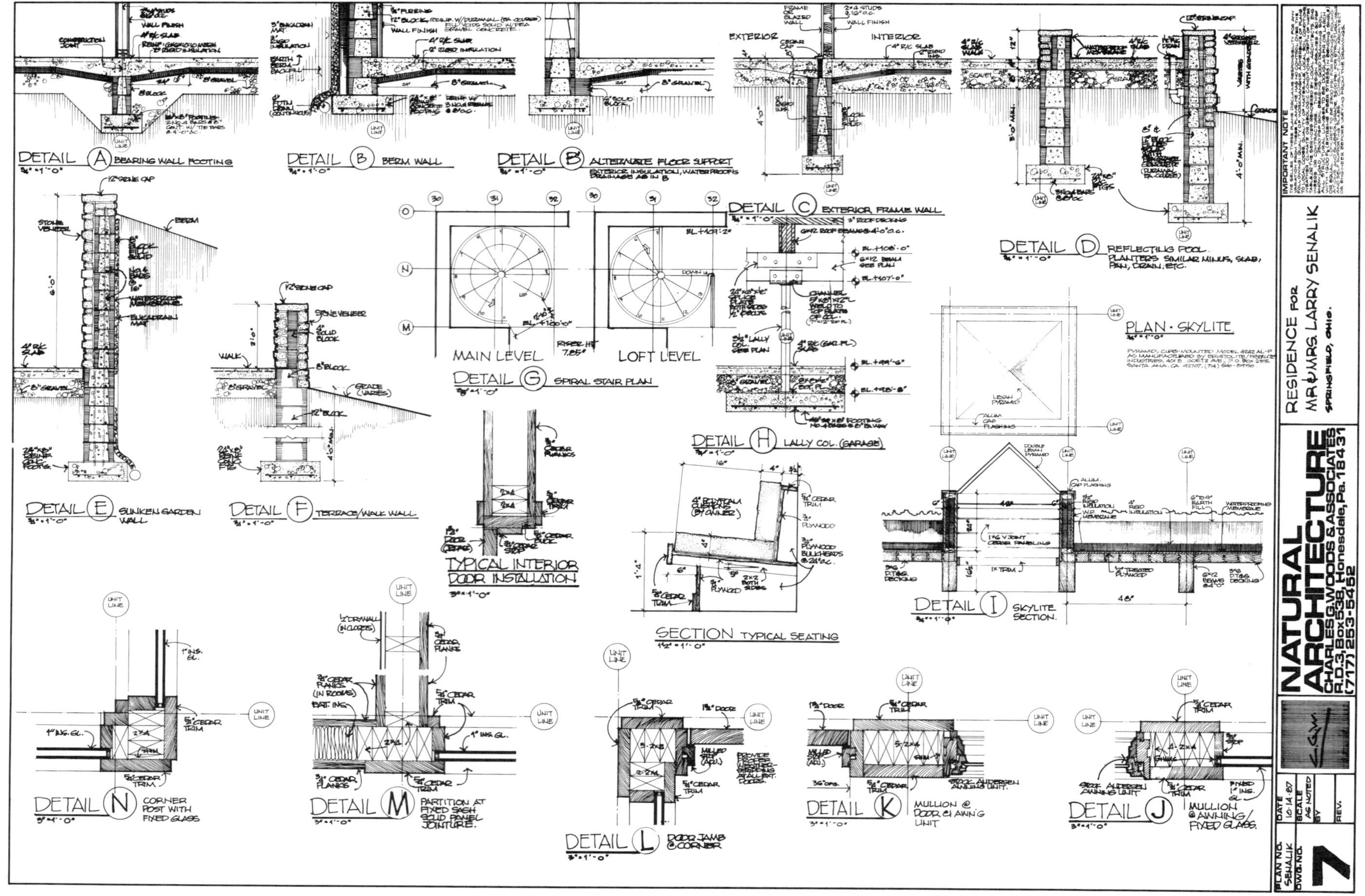
DETAIL A BEARING WALL FOOTING
DETAIL B BERM WALL
DETAIL B' ALTERNATE FLOOR SUPPORT
DETAIL C EXTERIOR FRAME WALL
DETAIL D REFLECTING POOL
DETAIL E SUNKEN GARDEN WALL
DETAIL F TERRACE/WALK WALL
MAIN LEVEL
LOFT LEVEL
DETAIL G SPIRAL STAIR PLAN
DETAIL H LALLY COL. (GARAGE)
PLAN · SKYLITE
DETAIL I SKYLITE SECTION.
TYPICAL INTERIOR DOOR INSTALLATION
SECTION TYPICAL SEATING
DETAIL N CORNER POST WITH FIXED GLASS
DETAIL M PARTITION AT FIXED SASH SOLID PANEL JOINTURE.
DETAIL L DOOR JAMB @ CORNER
DETAIL K MULLION @ DOOR & AWNING UNIT
DETAIL J MULLION @ AWNING/FIXED GLASS.
RESIDENCE FOR MR & MRS. LARRY SENALIK
SPRINGFIELD, OHIO.
NATURAL ARCHITECTURE
CHARLES G. WOODS & ASSOCIATES
R.D. 3, Box 538, Honesdale, Pa. 18431
(717) 253-5452
IMPORTANT NOTE
7

Senalik Residence construction plans.

A final thank-you and disclaimer* to Frank Lloyd Wright, by Charles

Much of what is good in this book goes back to Frank Lloyd Wright's tradition of architecture. I have studied his works and I have studied with two of his students, and know some of his closest students. Mac, too, has long admired his work.

But although there are in my work visual Wrightian influences from time to time, I have tried to imbibe the Master's "principles," not his "details", as some of his lesser followers have done at times.

Other great students of Wright like John Howe or the late great Wes Peters combined Wright's principles and details with real creative genius! But I know of no one else better suited (and there are many) that has attempted to do a primer on this tradition's house design, so I attempted it. How far I succeeded others will have to say.

To all who have helped me in my understanding: thanks, and to Frank Lloyd Wright a deep bow.

"I tried, Mr. Wright."

---

* Mr. Wright is not of course responsible for any errors or my own ugliness.

you don't have to apologize, Charles; you aren't that bad looking.

Woodsian Interlude.

Bermed T. .. as seen in Building Ideas magazine

AND NOW: MAC'S WORK

Dream house*

These are all versions of a house I'd like to build.

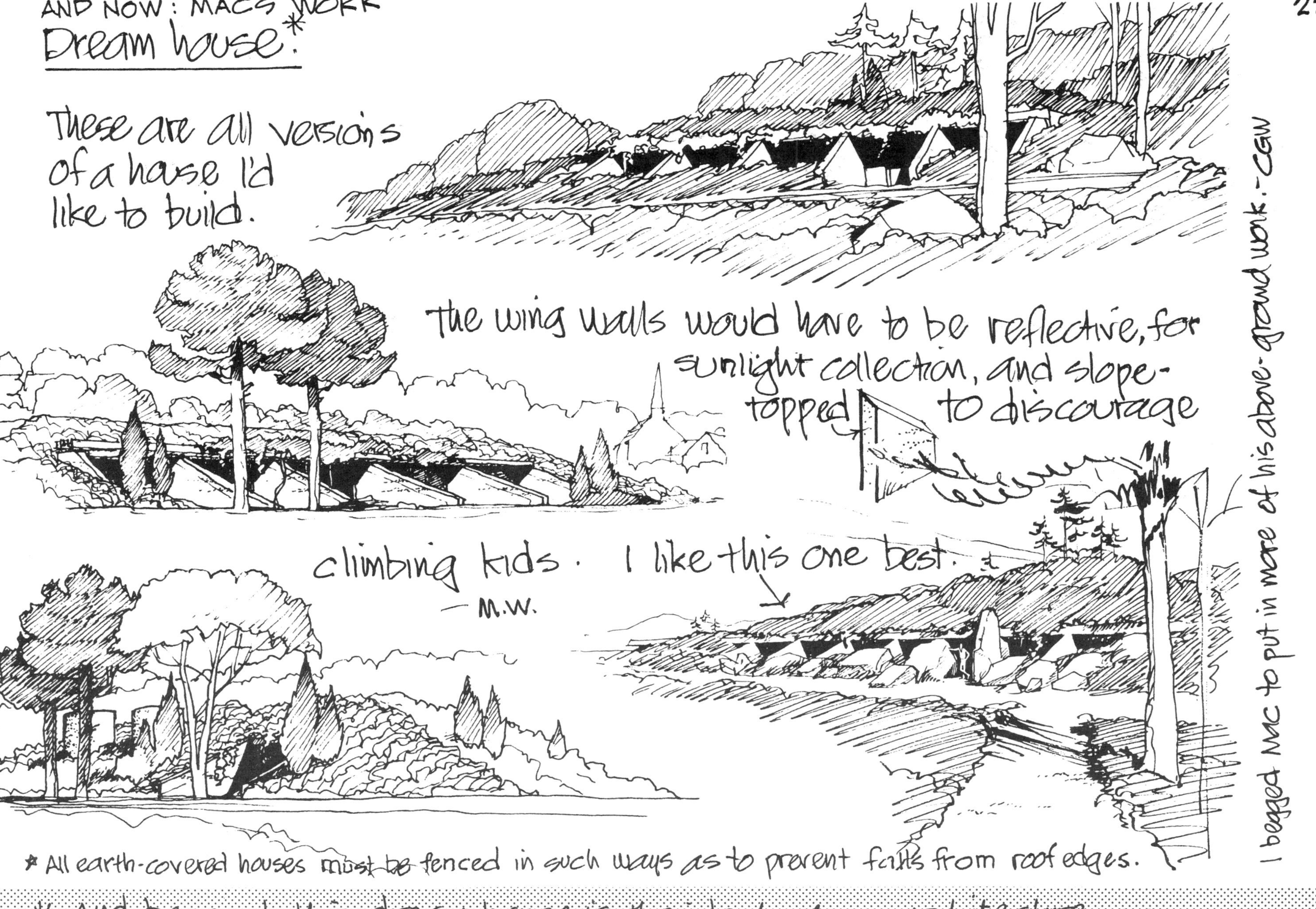

The wing walls would have to be reflective, for sunlight collection, and slope-topped to discourage climbing kids.

— M.W.

I like this one best.

I begged Mac to put in more of his above-ground work. – CGW

* All earth-covered houses must be fenced in such ways as to prevent falls from roof edges.

* And beyond this dream house is the ideal of an architecture so appropriate it almost disappears. Imagine America then!

A house by Mac

I did this little earth-covered house for *Popular Science* magazine. The design is reversible; it can be made to feature a northside

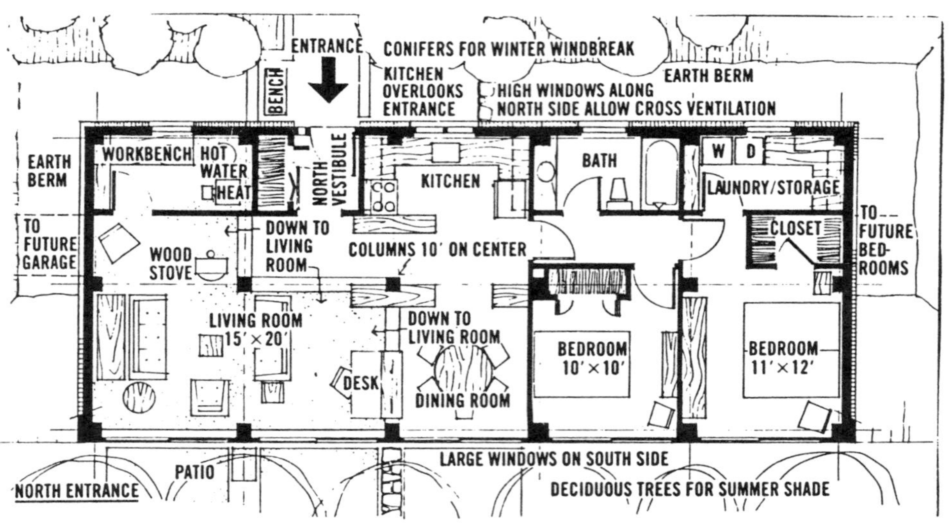

entrance as shown here or a southside entrance as shown on the next page.

construction plans & specs (for both houses in a single package) cost $95. Malcolm Wells Box 1149 / Brewster MA 02631. USA.)

Where groundwater level is no problem, earth shelters can be set deeper into the ground but even a ground-level floor will work well.

Only a little rearrangement was needed to turn the house into its south-entrance version. A passive solar building, it should perform quite well.

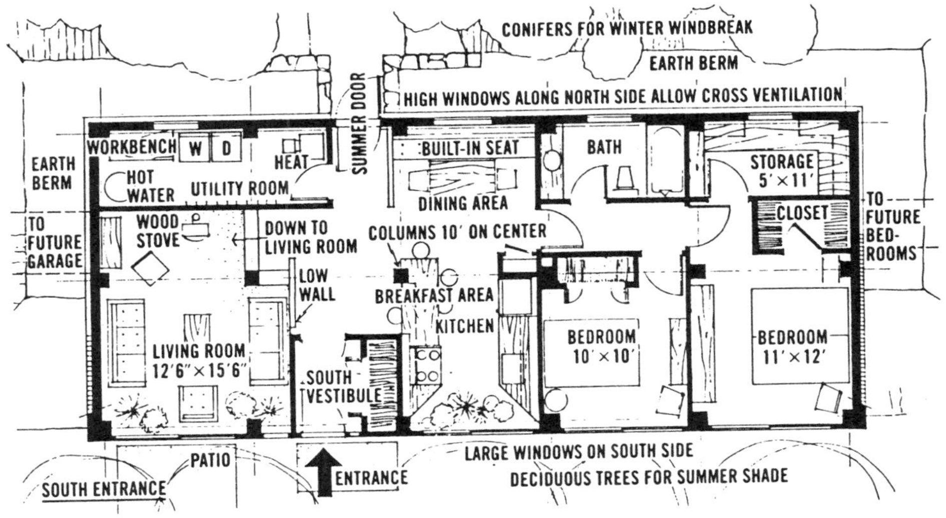

Be sure to modify this, or any temperate-climate, design before using it in the hottest parts of the country or Antarctica.

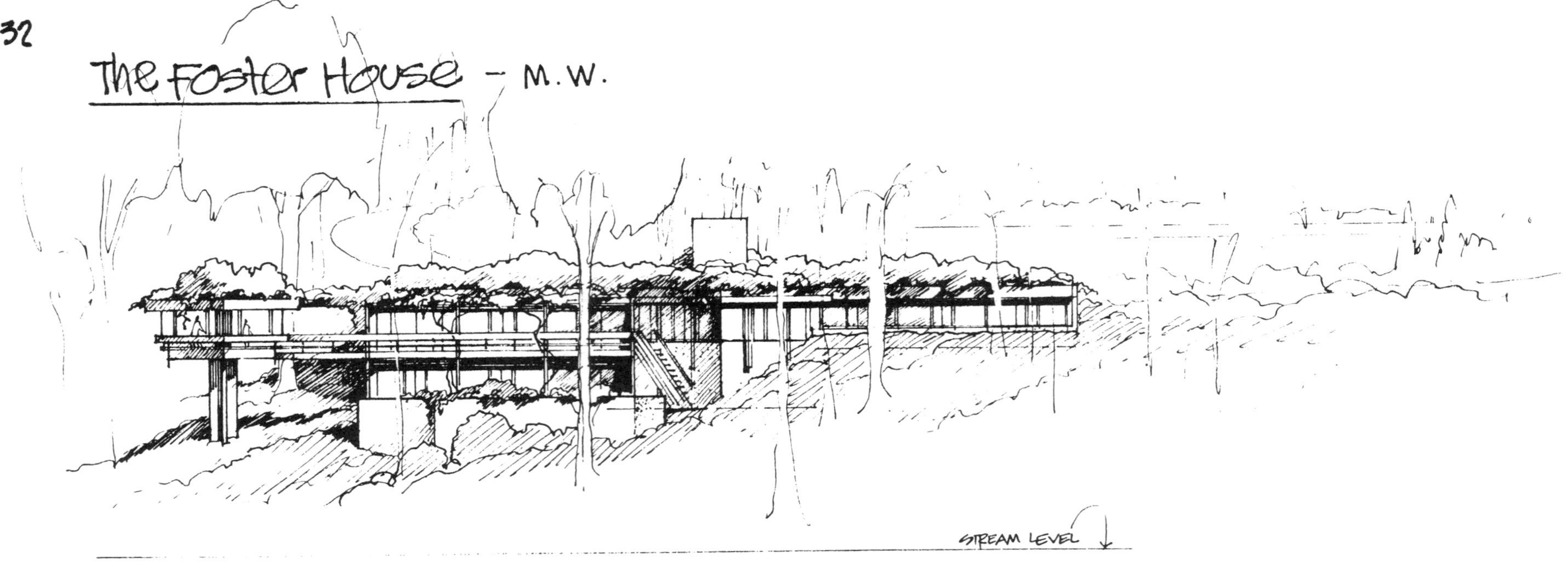

Many years ago, in the land of America, I was asked to do an earth-covered house on a wooded site above a beautiful stream. This was the south side, facing the valley. But then, as is the case with so many architects'

Horizontality in any building expresses the idea of water conservation and erosion control just as contour plowing does on farms.

The Foster House.

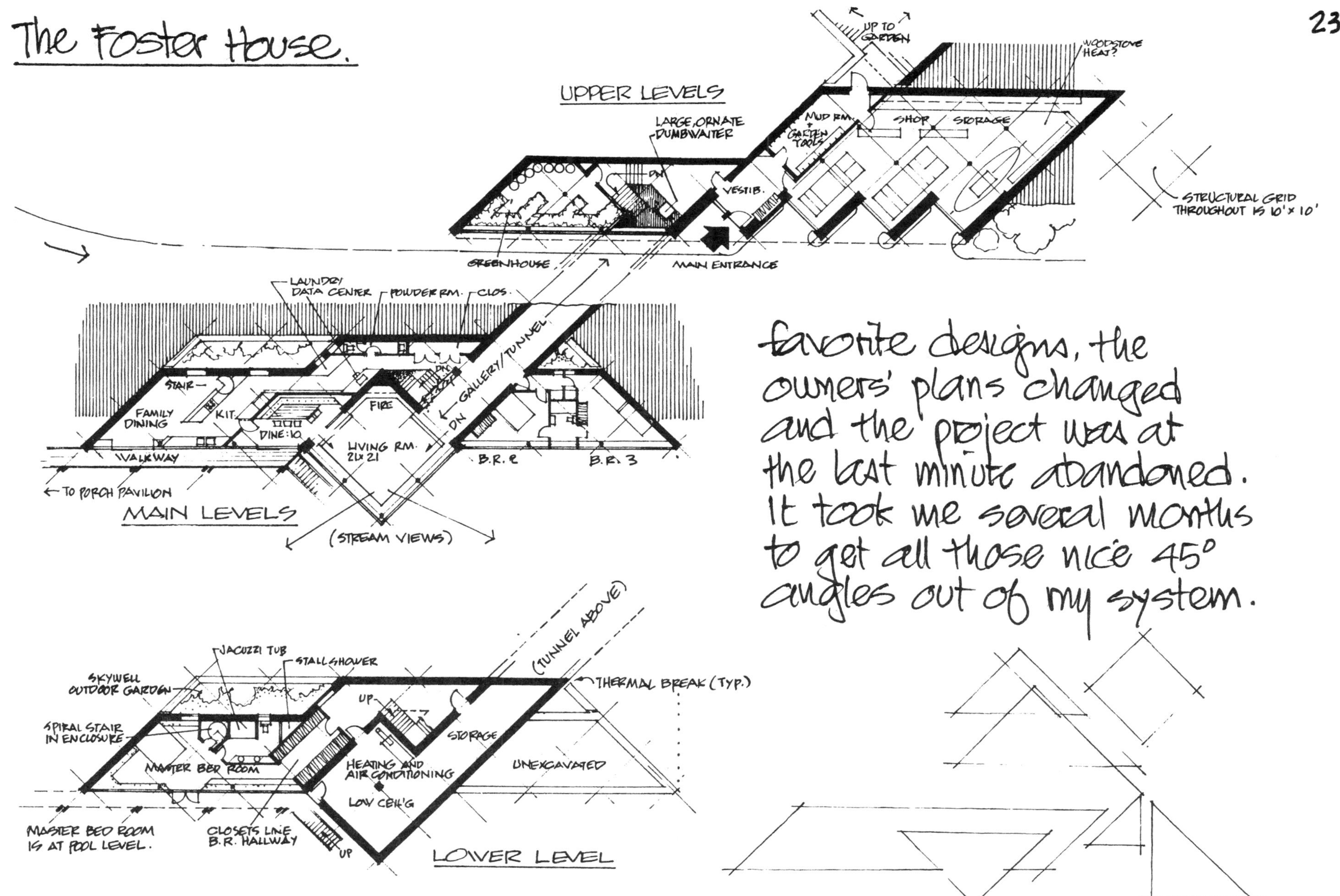

favorite designs, the owners' plans changed and the project was at the last minute abandoned. It took me several months to get all those nice 45° angles out of my system.

As should be the case with all floor plans, north is toward the top of the page. It helps the solar aspect of the design become more easily grasped.

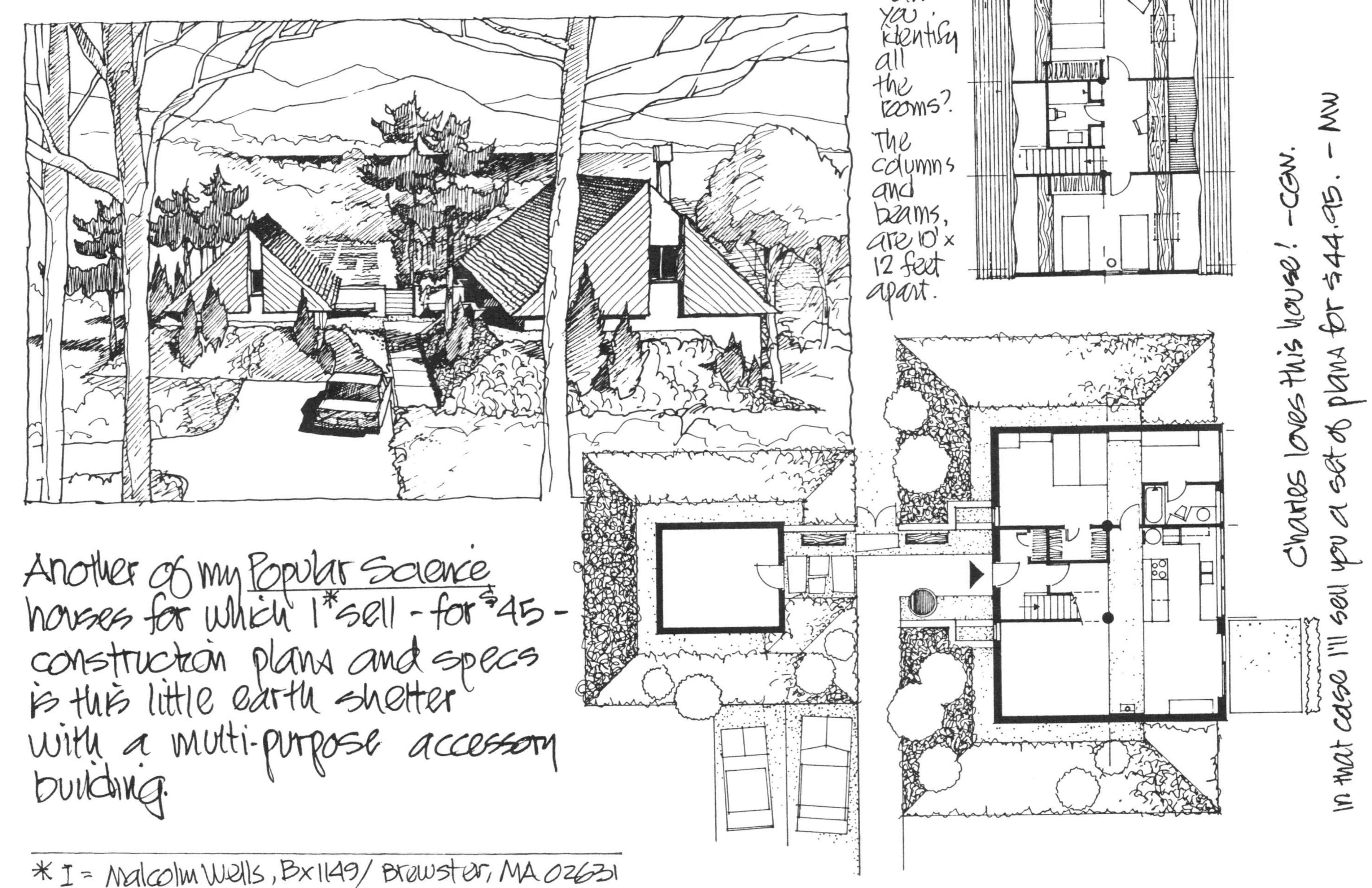

Can you identify all the rooms? The columns and beams are 10' x 12 feet apart.

Another of my Popular Science houses for which I* sell – for $45 – construction plans and specs is this little earth shelter with a multi-purpose accessory building.

Charles loves this house! —CGW.

In that case I'll sell you a set of plans for $44.95. —MW

* I = Malcolm Wells, Bx 1149/ Brewster, MA 02631

A small basement to house a human wastes composter would make this house much more benign environmentally. But be sure the structure is done right.

Los Angeles hilltop house. - M.W.

Another house that almost went ahead, this one was done for an open hilltop in the northernmost corner of sprawling Los Angeles County.

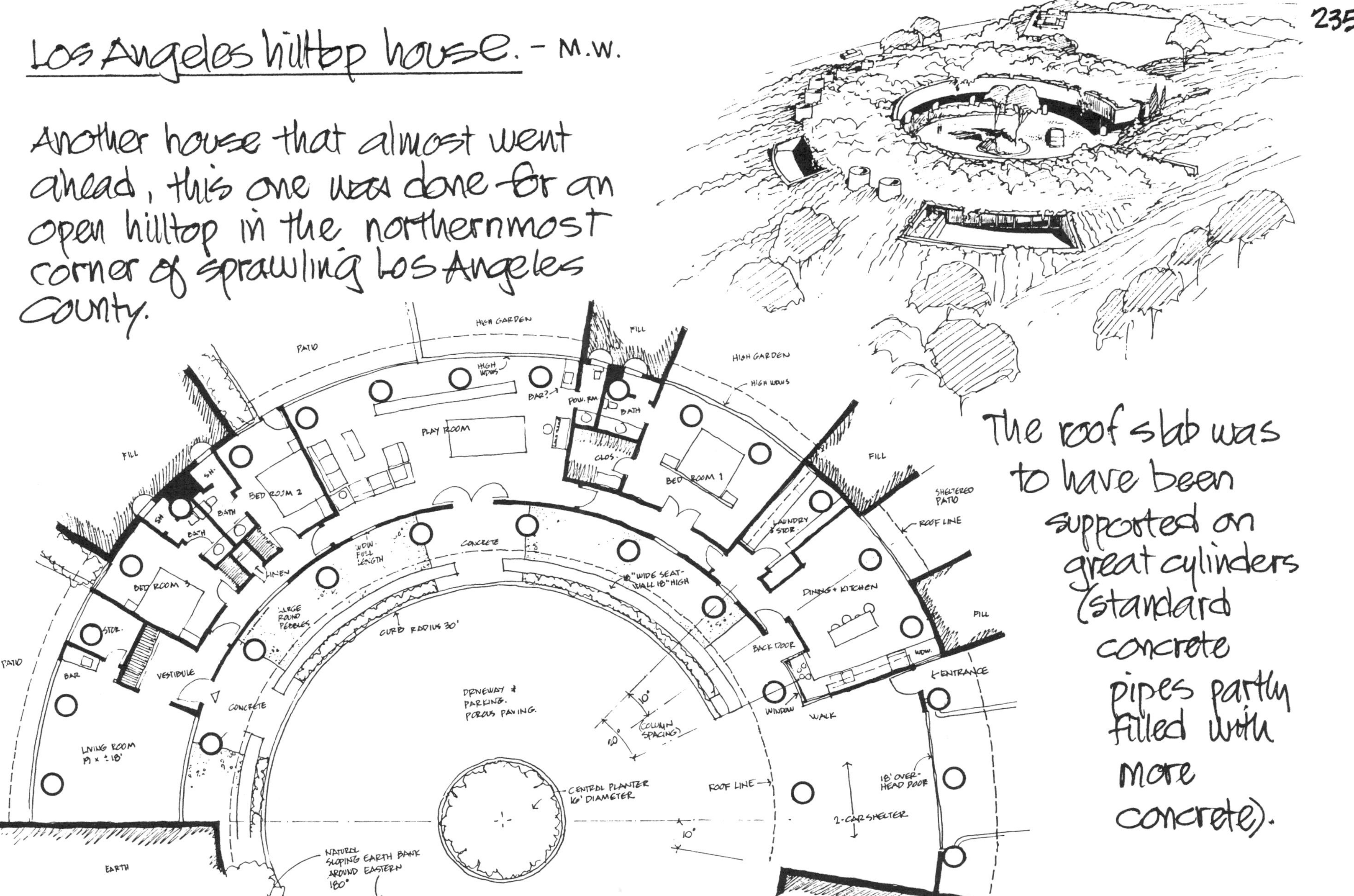

The roof slab was to have been supported on great cylinders (standard concrete pipes partly filled with more concrete).

As you can imagine, it's a different world from Cape Cod out there. Plenty of sunlight, mild winters, and — much of the time — water shortages.

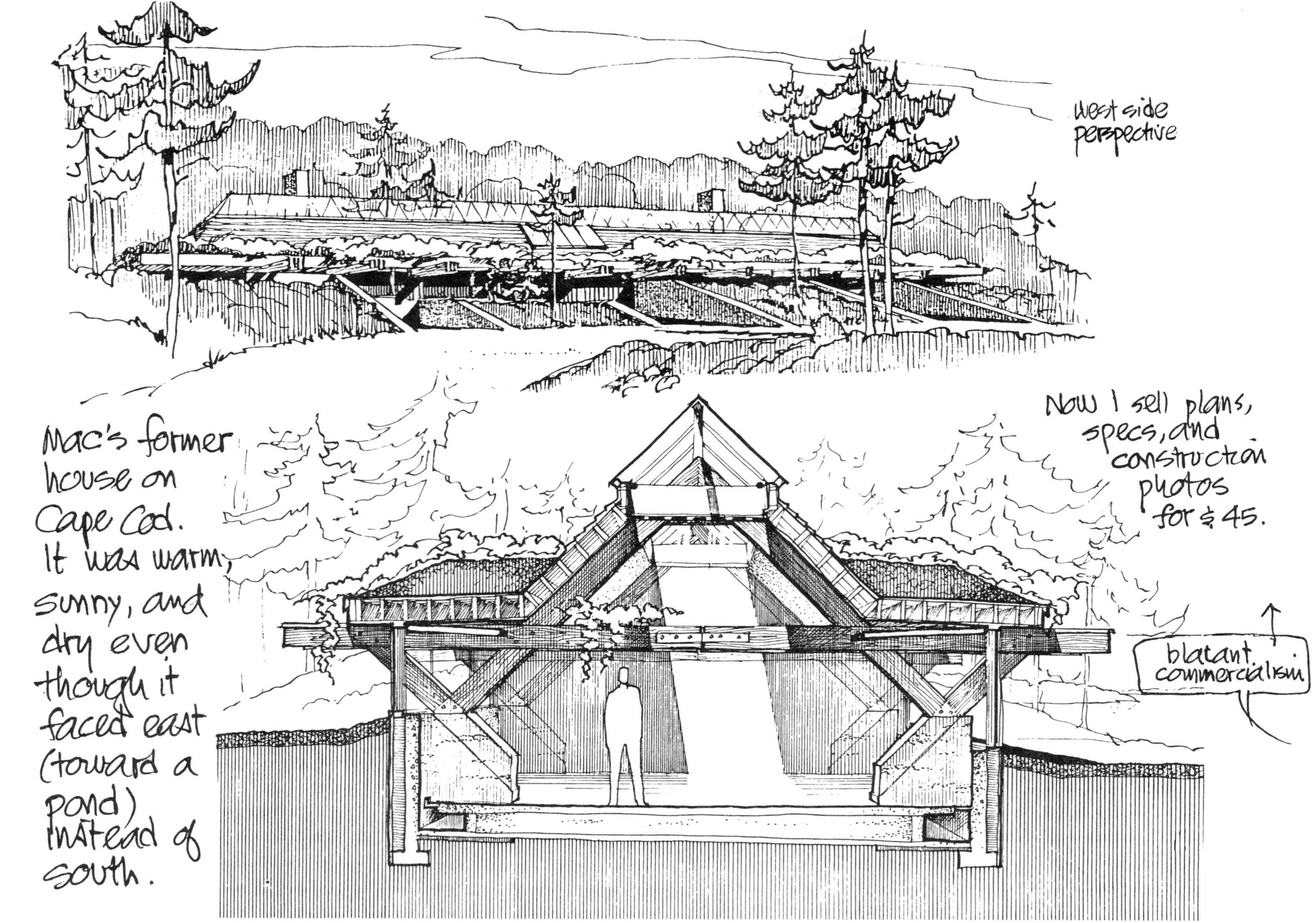

West side perspective
Mac's former house on Cape Cod. It was warm, sunny, and dry even though it faced east (toward a pond) instead of south.
Now I sell plans, specs, and construction photos for $45.
blatant commercialism

But climbing up into that triangle each fall to add the translucent insulating panels was no fun. I've thought of better arrangements.

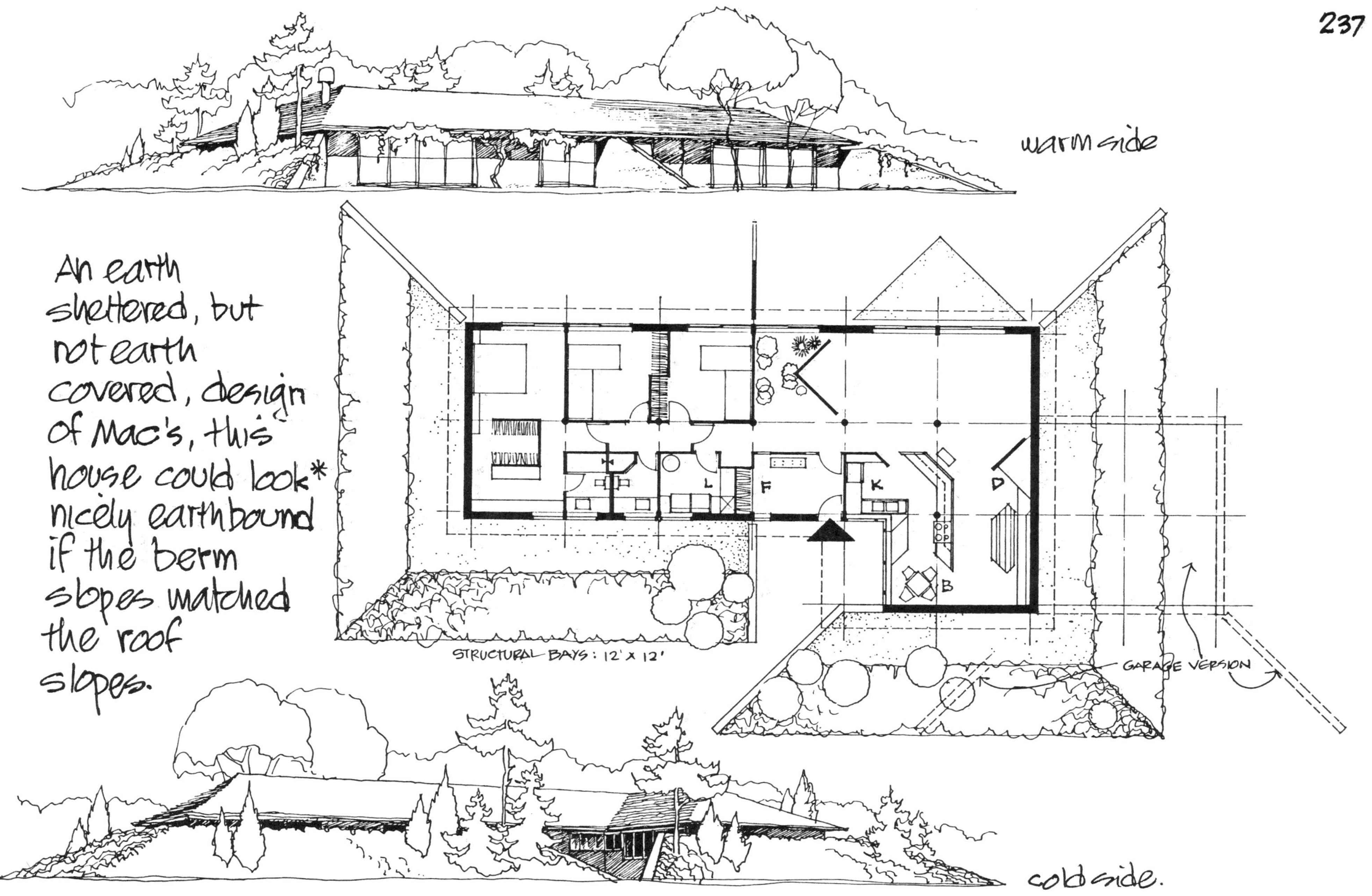

"I thought appearance was trivial in a time of planetary crisis." That was my conscience speaking, and she's right: I lost my head.

And a final word, from Mac: feature the plants, not the toys, in *your* landscape.

This page, dear reader, is not for you, for we know that you are a paragon of good taste as well as of design excellence. But, just in case someone known to have abominable taste happens to visit your neighborhood, you might want to leave this page open somewhere where he is likely to see it and be drawn to study these four appalling drawings.

If he does show an interest you might explain to him that mowed lawns are embarrassed enough by their mutilation without having the adornment of automobile tires, Mexican boys, flamingoes, deer, or windmills to explain to everyone.

Explain further that he can see plenty of such things in the great theme parks to which Americans flock in such large numbers.

KEEP AMERICA BEAUTIFUL

Thanks very much. And goodbye.

"Goodbye!" -Charles.

In our next book we hope to tackle the related subject of excessive and tasteless, not to mention wasteful, Christmas decorations and lighting. Bah.

Charles has very kindly agreed to shorten his 150-title
Suggested Reading List to fewer than twenty!

1. The Good House. Jacobson, Silverstein and Winslow. The Taunton Press. 1990 (Excellent!)
2. How to Build A House with An Architect. Baker, John, Milnes. Harper & Row. 1988. (Great!)
3. The Little House. Armstrong, Leslie. Collier. (Good!)
4. The Tao of Architecture. Chang, Amos. Princeton.
5. A Holy Tradition of Working. Gill, Eric. Phanes Press.
6. 30 Energy-Efficient Homes You Can Build. Wade, Alex. Rodale Press.
7. Gentle Architecture. Aw, come on, Charles, they've had enough of me. Wells. McGraw-Hill. 1990.
8. Natural Solar Architecture. Wright, David. Van Nostrand Reinhold.
9. The Natural House. Wright, Frank Lloyd. New American Library.
10. Natural Architecture. Woods, Charles. Element Books
11. The Complete Earth Sheltered House. Woods, Charles. Available from author.
12. Tao Te Ching. Lao Tzu.
13. Art and Scholasticism. Maritain, Jacques.
14. The Ethics. Spinoza, Benedictus.
15. The Summa Theological. St. Thomas Aquinas. Ed. McDermott. (1-vol. sel.) Christian Classics.
16. The Healthy House John Bower. Lyle Stuart 1989 (Very important!)

Index

Dear Reader,
You will no doubt by now have noticed Mac has said "Charles said this" a lot. It was not humility on his part. He is just thinking the reviewers will rip me to shreds and love his drawings! Ha ha.
—Charles.

I am moved more by senility than humility, Charles.

To Order Plans

(Charles's plans, that is; for Mac's you have to find him at Brewster, Mass., and beg.)

Plans include: perspective, plans, elevations, sections, details, and general specification, etc.

To order: please indicate name and page number of design. One set: $250, outside U.S.: $275 in U.S. funds, payable through a U.S. bank. Additional sets of same plan at purchase time: $35. Detached garage plans (where applicable): $35 each. Mailing and handling: $20 with each order.

NATURAL ARCHITECTURE
John Martin, Architect
Charles G. Woods, Design.
RD 3, Box 538 Honesdale, PA 18431
Office: (717) 253-5452
fax: same
Home: (717) 253-0891

Plan Name ____________________ Page ______
Plan Name ____________________ Page ______
Plan Name ____________________ Page ______
Extra Sets ____________________ Total costs $ ______

Name ____________________
Street Address ____________________
City ____________ State ______ Zip ______
Visa/M.C.
Credit Card No. ____________________ Exp. date ______
Signature ____________________
Phone Number ____________________

My two previous books, NATURAL ARCHITECTURE and THE COMPLETE EARTH SHELTERED HOUSE can be ordered from Natural Architecture for $20 each postpaid.